PLASMA
AT WORK

George A. Sites

BACKGROUND

THE PURPOSE OF THIS BOOK IS TO INTRODUCE THE READER TO MANY OF THE APPLICATIONS OF THE PLASMA ARC. THE APPLICATIONS ARE ALMOST ENDLESS AND IT IS, IN MY OPINION, ONE OF THE EMERGING TECHNOLOGIES OF TOMORROW. HOPEFULLY, IT WILL HELP READERS UNDERSTAND WHY PLASMA IS SO USEFUL. FOR THE ELECTRICAL ENGINEERS, SECTION 2 WILL HELP EXPLAIN THE DIFFERENCES BETWEEN THE DIFFERENT TYPES OF POWER SUPPLIES USED TO CONTROL PLASMA ARCS.

George A. Sites

FROM THE AUTHOR

I DESIGNED MY FIRST PLASMA POWER SUPPLY IN 1978 FOR WESTINGHOUSE PLASMA. THE APPROACH USED ON THAT POWER SUPPLY HAS BEEN MODIFIED SLIGHTLY OVER THE YEARS BUT IS STILL EFFECTIVE AND IN USE TODAY BY MANY OF MY CLIENTS. IN ASSOCIATION WITH OTHER ENGINEERS AND CLIENTS, I HAVE USED OTHER CIRCUIT TOPOLOGIES WITH GREAT SUCCESS ALSO. ALTHOUGH I HAVE DESIGNED AND BUILT PLASMA POWER SUPPLIES FOR MANY CLIENTS OVER THE YEARS, WESTINGHOUSE PLASMA STILL REMAINS ONE OF MY CLIENTS TODAY, NEARLY 30 YEARS LATER.

DEDICATION, THANKS and IN MEMORY OF

I dedicate this book to my parents, George and Frances Sites for all that I am today. As a young boy, not only did they teach me right from wrong but they taught me that it really doesn't matter what job or position I hold in life as long as I enjoy it, work hard, do my best to be the very best at whatever I do and of course, its honorable. I've worked hard to live up to their beliefs and standards and I know they are very proud of me.

I also dedicated this book in the loving memory of my Mother, Frances Josephine Sites who passed away June 30, 2004. I miss her and she will be forever in my heart and memories.

NOTES

Unless otherwise indicated, all drawings are by
Charles E. Denney, Jr. or Tom Forbus

Edited by Alida L. Breen
& Roy A. Sites, M.L.A.

Covers Designed by Wayne L. Cox
& Roy A. Sites, M.L.A.

Other Books by George A. Sites

"An Introduction to SCR Power Controls"

Available at:

www.amazon.com

www.cafepress.com/powercontroller

Published Articles by George A. Sites

"Selecting an Enclosure for SCR Power Controls"
Industrial Heating – May 1991

"Plasma Arc Heating – An Overview"
Industrial Heating – November 1994

**"Controlling Silicon Carbide Heating Elements
with SCR Power Controls"**
Industrial Heating – September 1997

"Understanding SCR Power Controls"
Process Heating – March 1999

"Improving Plasma Spray Coating Control"
Process Heating – March 2004

"Forehearth Control for Electric Glassmaking"
Glass Magazine – April 2005

"Utilizing Multi-Zone Power Control Systems"
Process Heating – May 2007

ISBN-13: 978-1500169206

ISBN–10: 150016920X

Table of Contents

SECTION 1 – PLASMA AT WORK

SECTION 2 – PLASMA POWER SUPPLIES

LIST OF PHOTOGRAPHS AND ILLUSTRATIONS

SECTION ONE

PLASMA AT WORK

Chapter 1

WHAT IS PLASMA?

INTRODUCTION

Until recently, heat processing was typically limited to conventional fossil fuel combustion, electric resistance heating and/or induction heating. Modern advancements in heating technologies have created more efficient options. In my opinion, one of the most exciting is the Plasma Arc Heating System especially when used for waste incineration.

Plasma energy is a common, naturally occurring resource. The lightning in a thunderstorm is a perfect example. Another example is the clear globes seen in specialty shops with blue or purple arcs emanating from a ball in the center and out to the outer glass skin.

Originally used in the late 1800's for manufacturing cement and fertilizer, plasma applications have proliferated. Numerous applications include the following:

- tundish heating
- ferrous and non-ferrous melting
- hazardous waste treatment and remediation
- industrial diamond manufacturing
- plasma spray coating
- steel cutting
- soil remediation/vitrification
- gasification of wastes including municipal solid waste

In the early 1960's, NASA used plasma heating to simulate the extremely high temperatures of re-entry into the earth's atmosphere. Although the applications are widespread, plasma is just becoming recognized as one of the technologies of tomorrow, especially for waste treatment.

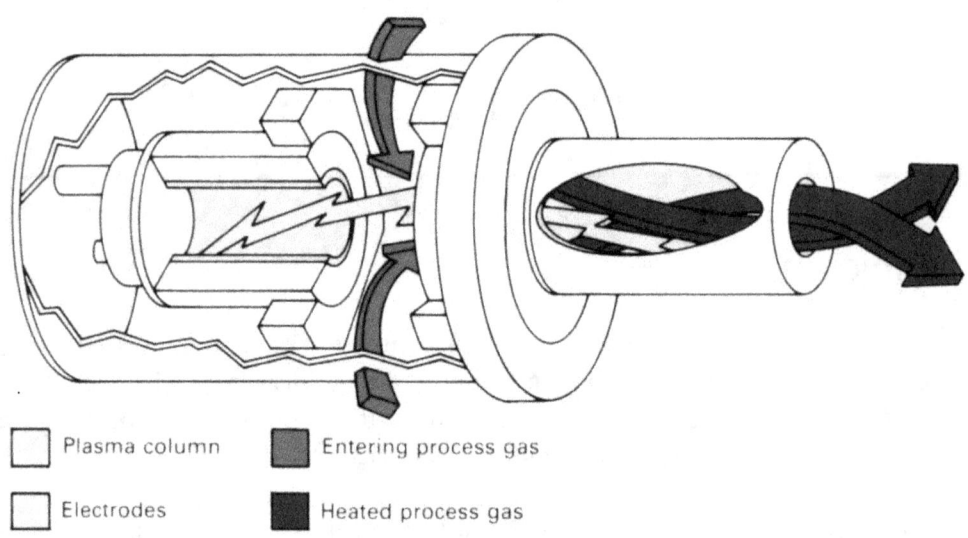

| | Plasma column | | Entering process gas |
| | Electrodes | | Heated process gas |

Figure 1 – Cutaway of a typical Plasma Torch
(Diagram courtesy of Westinghouse Plasma Corp.)

ADVANTAGES

Advantages of the plasma heating system are the following:

- greater temperature control
- faster reaction times
- better processing control
- lower capital costs
- greater throughput
- higher temperatures
- more efficient use of energy as compared to the more conventional methods.

The plasma process is very efficient. Very quickly and with very little combustion, temperatures up to 20,000 degrees Fahrenheit and higher can be achieved – hotter than the surface of the sun. Since very little combustion takes place, stack gasses are virtually eliminated minimizing environmental concerns.

Plasma arc technology can create plasma using nearly any type of gas. For example, gasses such as oxygen, helium, nitrogen, air and carbon monoxide can be used. One of the greatest advantages of plasma in process heating is that it contains more available energy for work than fossil fuel flames at high temperature. For example, at about 2900F using fossil fuel only about 20% of the energy is available for melting metal in fossil fuel heating as compared to more than 80% for plasma treated air.

The only drawback/negative is that OSHA has been slow to approve plasma heating as a safe and viable alternative method for waste disposal here in the United States. Tests by many plasma torch system manufacturers and universities have shown that plasma is definitely a suitable alternative method for high temperature heating and should be considered.

SUMMARY

The benefits of plasma far outweigh the disadvantages and the plasma process has been proven to be a safe and reliable high temperature heat source. The plasma heating process is being used all over the world for numerous applications. Of the many conventional methods used today to create heat in a process, most are typically limited to the combustion temperature of the associated fuel. By having the unique capability of increasing the energy of the process gas two to ten times higher than conventional combustion equipment, the plasma technology offers many distinct advantages over traditional methods.

NOTES CHAPTER 1

Chapter 2

COMPLETE PLASMA HEATING SYSTEM

INTRODUCTION

The complete plasma heating system is made up of many components. Each of these can vary from one plasma heating system to another. The most common components include:

- Plasma Torch
- DC Power Supply
- Furnace
- Control System
- Gas Supply System
- Cooling Water System

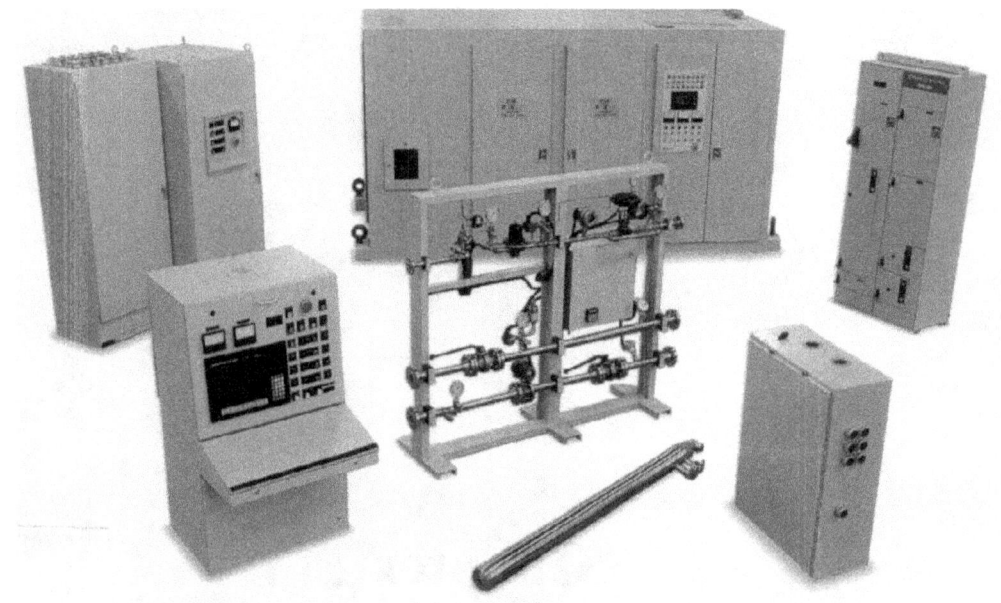

Figure 2 – Components of a Plasma Heating System
(Photo courtesy of Plasma Energy Corporation)

PLASMA TORCH

The plasma torch is the device that delivers the heat to the load regardless whether it is a combustion chamber in a furnace, a type of waste needing disposed of or a cutting application. Two basic types of torch designs are used - transferred and non- transferred. Each has its own specific uses.

The transferred torch initiates an arc internal to the plasma torch and then transfers the arc to a work piece such as steel for cutting or municipal waste for incineration/remediation.

The non-transferred torch initiates an arc internal to the plasma torch where the arc remains. Gas/air pressure through the torch cause the arc to bend or stretch out of the torch housing. See Figure 3.

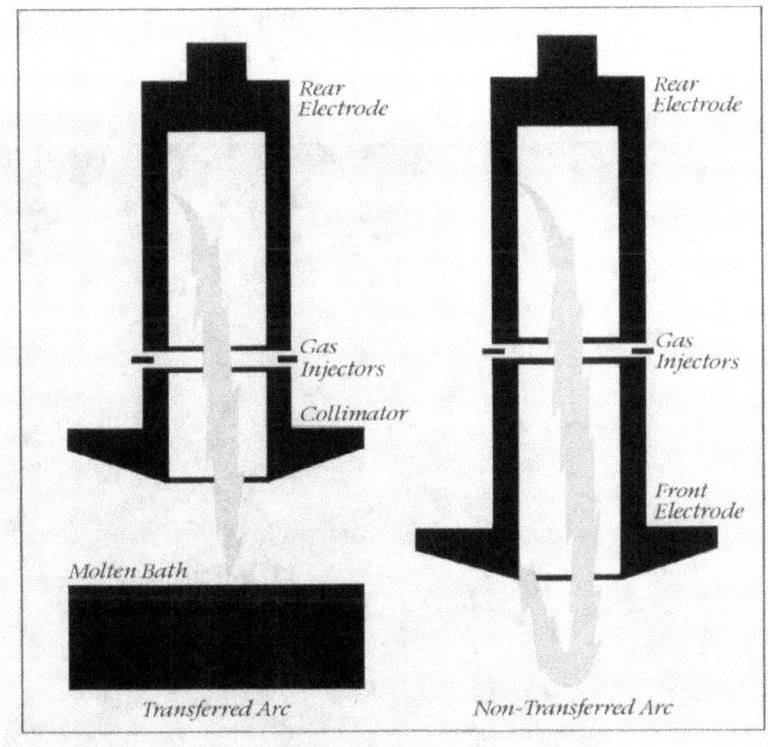

Figure 3 – Transferred and non-transferred Torches
(Diagram courtesy of Plasma Energy Corporation)

The plasma torch is loosely based on jet engine technology and is complex in design.

DC POWER SUPPLY

The power supply is the heart of the overall power control system. Specially designed electronic circuitry provides extremely stable arc current during ever changing load conditions. Depending on the exact application, the more stable the arc, the more finely tuned the process.

Power Supplies range in power ratings from a few kilowatts to many megawatts depending on process requirements and variables.

An early example of a plasma power supply would be a typical current regulated welding supply. Each of the different power supplies are explained in detail in later chapters.

Figure 4 – Typical Welding Power Supply
(Photo courtesy of Lincoln Electric)

FURNACE

The furnace is the containment area where the plasma torch actually performs its work. This is true regardless of whether it is being used for waste disposal, steel melting, etc. The one exception is when it is used for steel cutting.

The furnace can be something simple in which the plasma torch is super heating the air as in a wind tunnel or something more complex such as a Cupola furnace used for steel making. See Figure 5.

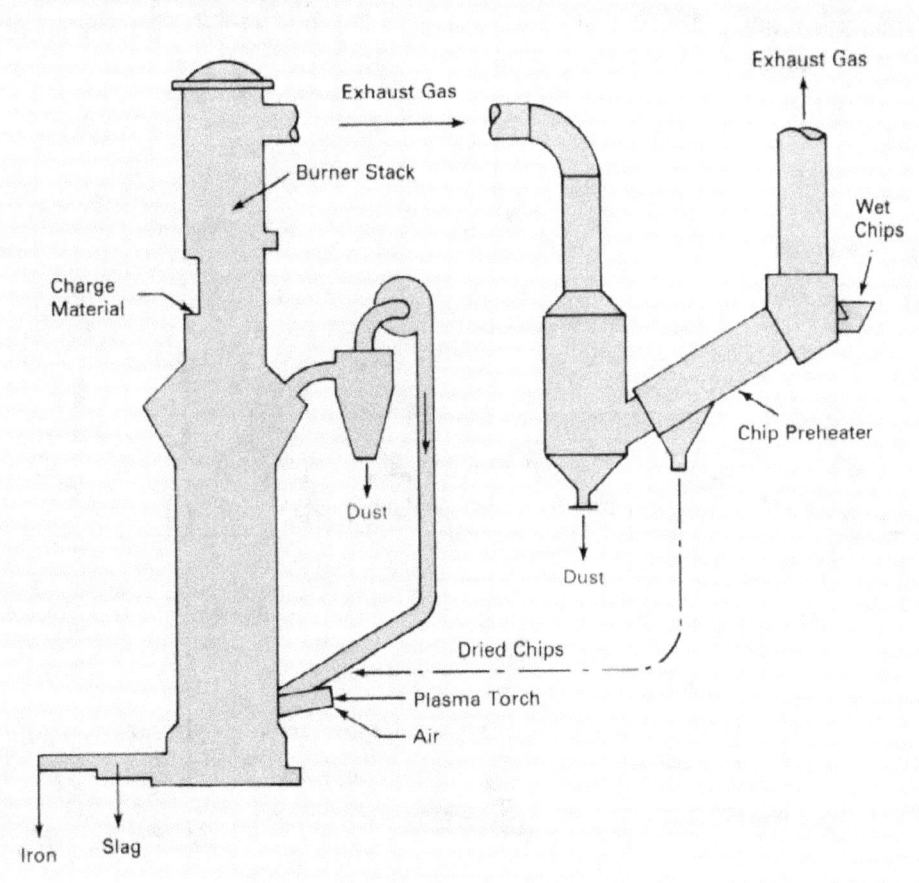

Figure 5 – Cupola Furnace
(Diagram courtesy of Westinghouse Plasma Corp.)

CONTROL SYSTEM

The control system provides the inputs to the power supply and the outputs to the other process equipment such as the gas supply system. It also monitors process variables such as arc current and voltage, gas flow and pressure, cooling water flow and other variables as required. This is done using a programmable logic controller (PLC) or other type of computerized equipment. It usually includes a monitor and other periphery equipment as required.

GAS SUPPLY SYSTEM

The gas supply system supplies the various process gasses in the proper flow, pressure and time periods required by the plasma process. The torch manufacturer usually supplies the gas system because coordination of the gasses, flow rates and pressures is essential to proper torch operation.

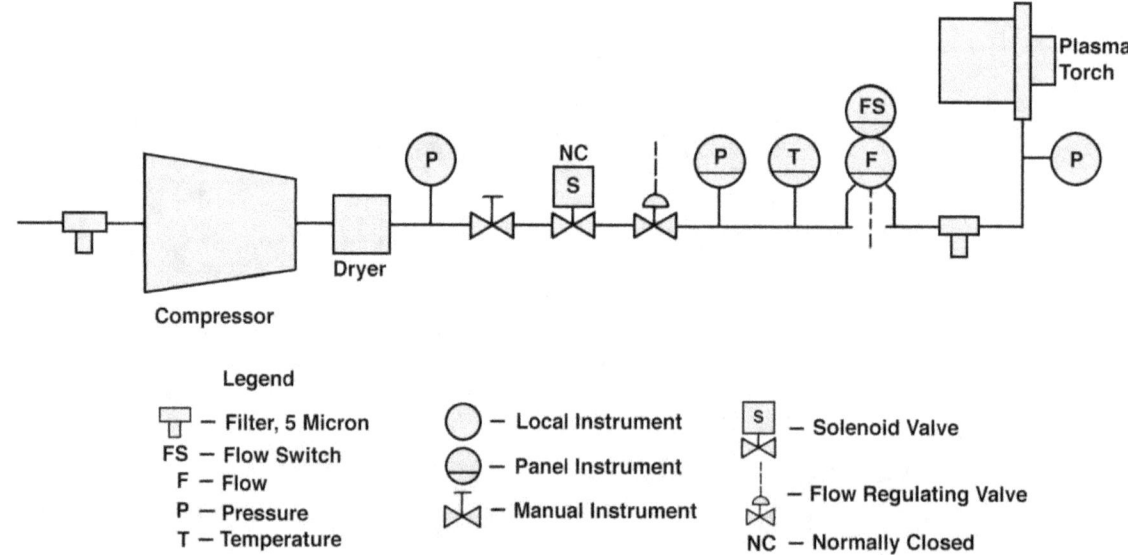

Figure 6 – Outline of a typical plasma torch gas system
(Diagram courtesy of Westinghouse Plasma Corp.)

A variety of reducing, oxidizing and inert gases can be used as the plasma torch process gas. Gas flow rates will dependon the gas composition and process requirements. An outline of a typical gas system is shown in Figure 6. The gas system usually includes a particle filter, flow regulation and measurement with low gas-flow alarms and also pressure/temperature indications.

COOLING WATER SYSTEM

The cooling water system supplies the cooling water for the DC power supply, the plasma torch and oftentimes the furnace. Not all power supplies or furnaces require cooling water but the plasma torch always does. The water system is crucial to the reliability of the complete plasma heating system. Loss of water can mean disaster if proper interlocks and shutdowns are not provided. Maintaining reasonable operating temperatures on the plasma torch reduces the electrode wear. The cooling water system is usually closed loop so maintenance is extremely important.

Typical plasma torch cooling water specifications are as follows:

- pH 9-9.9
- Resistivity 100,000 to 500,000 ohms/cm at 77F
- Dissolved oxygen 0.003 ppm
- 100-mesh (150 micron) strainer

WP20

Figure 7 - Typical Closed Loop Water Cooling System
(Photo courtesy of Water Saver Systems, Inc)

SUMMARY

The complete plasma heating system is fairly complex. However it can perform heating applications not normally available using fossil fuels. The plasma heating is very efficient and delivers much more of the energy as useful heat compared to conventional heating methods.

Chapter 3

PLASMA at WORK

INTRODUCTION

Industries and applications are many and wide-ranging. Although the plasma equipment is similar in most cases, the process designer must still specify the plasma operating parameters thereby making the equipment custom in nature. Some of the applications include the following:

- Municipal, Hazardous and Medical Waste Remediation
- Incineration & Gasification
- Boiler Ignition
- Foundry Melters
- R & D Applications
- Diamond Making
- Glass & Ceramic Coating
- Metal Cutting

•   These descriptions are intentionally brief in detail and are only intended to give the reader an idea of the capability of the plasma heating system. Many of these applications are covered in detail in the next few chapters in which actual applications are explained.

MUNICIPAL, HAZARDOUS and MEDICAL WASTE REMEDIATION INCINERATION/GASIFICATION

New challenges for plasma incineration/remediation exist today. These include low-level radioactive waste; soil and ground water contaminated by pesticides; and hazardous medical/biological waste.

Figure 8 – Plasma Torch Vitrifying Inorganic Waste
(Photo courtesy of Phoenix Solutions Co.)

The high temperatures of the plasma heating system significantly reduce the volume of waste. All that remains is an inert, non-leachable slag. In some cases it provides the opportunity to recover and recycle metals from the process. High destruction removal efficiencies in excess of 99.99999% have been achieved.

BOILER IGNITION

As we all know, there are many electric utilities that use coal fired boilers to create steam and in turn, generate electricity.

These boilers are shut down periodically and then fired back up, as demand requires. In most electric utilities, gas is used to create a flame and in turn ignite the coal. By replacing the gas igniter with a plasma torch, the coal ignites more quickly and the utility utilizes their own electricity rather than a fossil fuel they purchase from another source. In addition, the plasma torch is much more efficient.

FOUNDRY MELTERS

In the early 1980's, plasma heating systems were uncommon sights in foundries. Low cost fossil fuels and relaxed environmental regulations hindered the use of electro-technology. Ten years

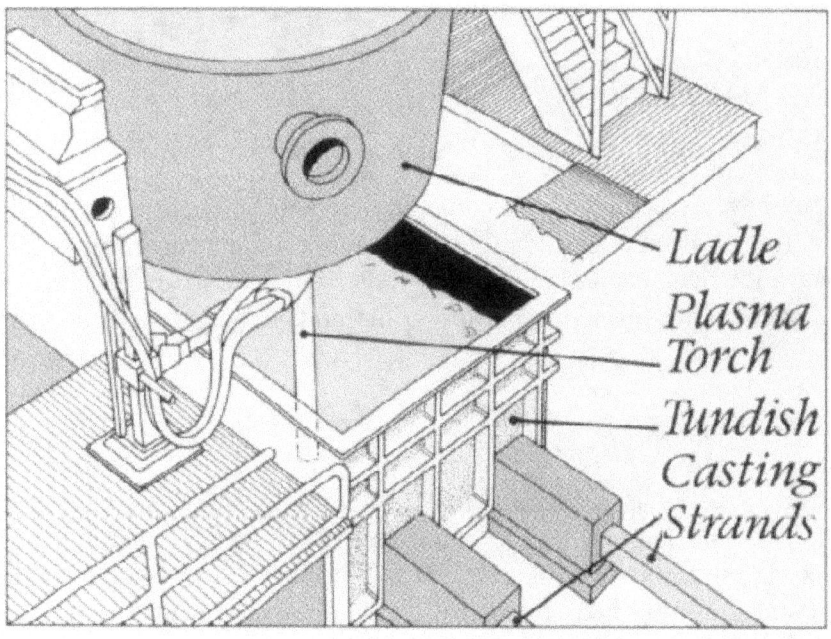

Figure 9 – Tundish Heating
(Diagram courtesy of Phoenix Solutions Co.)

later, foundries facing stricter regulatory enforcement and increasing environmental control costs started taking a fresh look at plasma heating systems as a source of high temperature process heat.

Two applications gained widespread acceptance and recognition in the United States and abroad in the Pacific Rim nations and Europe. The first used plasma to heat the Tundish in a steel factory; the second used plasma to heat the foundry Cupola during melting operations. An example of a Cupola Furnace is shown in Figure 5.

R & D APPLICATIONS

These applications are almost unlimited. For example, NASA has used them for simulating temperatures present in re-entry into the earth's atmosphere and other very exotic projects. Many of the National Labs use plasma for various experiments. Several universities that have metallurgical studies departments use plasma for testing the strength of different materials.

DIAMOND MAKING

It is well known that diamonds are a girl's best friend. But as one of the hardest materials known, they have other very significant characteristics. The Integrated Circuit industry uses diamonds for electronic applications because of their superior heat conductivity and exceptional electrical insulation ratings. Also diamonds can be used to coat the cutting surface of tools. The plasma torch allows diamonds to be grown or deposited very rapidly.

PLASMA SPRAYING

Plasma spray coating is a process that of thermally spraying one substance (usually in powder form) onto another. One example is coating architectural glass and another is coating jet engine turbine tip blades. There are many glass coating applications. The one that first comes to mind is architectural glass. The colored windows on a modern building such as a skyscraper are good examples of plasma coated glass. In this application, rather than gas, a material such as copper or aluminum is sprayed through the nozzle of the plasma torch. The high temperature of the plasma melts the material and sprays in onto the glass surface. The more stable the arc current, the more uniform or smooth the coating.

High temperature ceramic materials can be plasma sprayed onto a metallic surface such as an aircraft engine turbine tip blade. Turbine tip blades are spray coated with a high temperature ceramic coating that ultimately serves as the turbine's bearing surface.

Figure 10 – Examples of Turbine Tip Blades
(Photo courtesy of TTL)

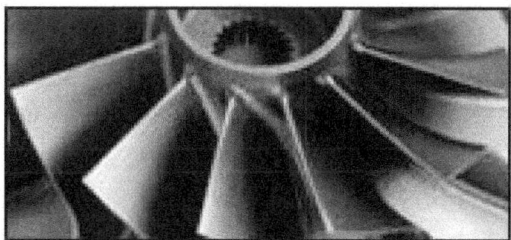 

METAL CUTTING

The use of a transferred plasma torch for cutting steel is a fast and exceptionally precise method. Steel thicknesses of over 1 inch can be cut like butter.

Figure 11 – Example of Plasma Cutting

Computerized positioning allows the user to cut intricate shapes such as those shown in Figure 11. As in other types of plasma applications, the more stable the arc current, the more precise the cut will be.

SUMMARY

Although the plasma torch has been around for a long time, the applications are just becoming known. Now with advancements in power electronics and control methods, the applications are more widespread than ever. This is due to the increased controllability of the arc and the rising cost of fossil fuels.

The small size, minimal installation costs and the ability to use different gasses are all benefits of the plasma torch. However, the greatest advantage is the ability to provide and maintain accurate high temperature control.

NOTES CHAPTER 3

Chapter 4

SAVING THE ENVIRONMENT

INTRODUCTION

Chapter 4 presents the plasma process at work helping the environment. In my opinion, these applications have the most promise for the future of our world. This entire chapter is about actual working installations and/or experiments for waste incineration. All information was found on the internet and as such, is in the public domain or was provided by my clients to whom I am grateful. Much of information is about Westinghouse Plasma Corporation's projects because they have been the most willing to share information. However, many other Plasma companies such as Plasma Energy Corporation (PEC) and Phoenix Solutions are involved in similar projects.

MUNICIPLE SOLID WASTE (MSW)

As we all know, municipal solid waste or MSW is a serious problem throughout the world due to a shortage of landfill space. In fact, many cities in the United States have been hauling trash to other states using trucks, boats and/or trains.

According to Shyam Dighe, chief executive of Westinghouse Plasma Corporation (WPC), municipal dumps in the U.S. spend on average $30 a ton to dispose of trash. Plasma incinerators can cost up to $20 million raising the cost of incineration to $50 or as much as $80 per ton. However, when the value of recycled waste fuel is factored in, the cost per ton decreases.

Westinghouse Plasma has designed and built many plasma reactors for disposing of MSW. Three of these plants have been installed in Japan where there is a serious shortage of landfill space. A pilot plant was built and started up in Yoshii, Japan during 1999 and the results have been very good. Emissions are very much reduced and the slag is a glassy product that can be sold as an aggregate for roadbeds and other construction uses. Dioxins were measured at levels approximately 100 times lower than from a conventional incineration plant. For organic waste, the production of power via a combustion turbine/combined cycle at much higher efficiencies (approximately 40% thermal efficiency versus approximately 20% for an incineration steam boiler plant) is an added benefit which makes the project cost effective compared to conventional incinerator/steam boiler MSW plants. Additionally, the need and expense to dispose of ash is eliminated.

A second plant was put into operation in 2002 and turned over for commercial operation in April 2003. This plant was designed by Hitachi Metals and utilizes Westinghouse Plasma torches, power supplies and gas systems. Located in Utashinai, Japan it was designed to primarily use automobile shredder residue (ASR) as fuel at approximately 165 tons per day (tpd) capacity but was also designed to operate with a 50% mixture of MSW. It can also process up to 300 tpd of 100% MSW. While incinerating, it produces approximately 8 MW of electric power.

A third plant built by Hitachi Metals and Westinghouse Plasma is located in the twin cities of Mihama and Mikata, Japan. It processes 24 tpd of MSW and 4 tpd of sewage sludge.

More information can be found on each of these installations at the following websites:

http://www.hitachi-metals.co.jp

http://www.weshinghouse-plasma.com

Phoenix Solutions Company has Plasma incinerators installed in Mita and Honjo, Japan. Both are used for MSW incineration and generate electricity during incineration that is sold back to the municipality. More information on Phoenix Solutions can be found at:

http://www.phoenixsolutionscompany.com

PYROLYSIS

According to experts in the plasma industry, pyrolysis is defined as the process of causing a chemical change through heat in an oxygen-starved environment.

One of the major concerns about the treatment and subsequent disposal of chemical wastes is the cost and the liability associated with transportation and disposal. The use of the Westinghouse Plasma Mobile Pyroplasma Incinerator helps eliminate these concerns by destroying the chemical wastes right on site.

The entire plasma system is housed in a standard 48 foot trailer ready to go to work. All that is needed on site is an AC power source, a water supply and a sanitary sewer system. All other hardware and process control equipment are contained within the trailer. The plasma torch that destroys the waste, the reaction chamber for forming non-hazardous compounds and other process equipment is located in the center of the trailer. A fully equipped laboratory is in the front of the trailer.

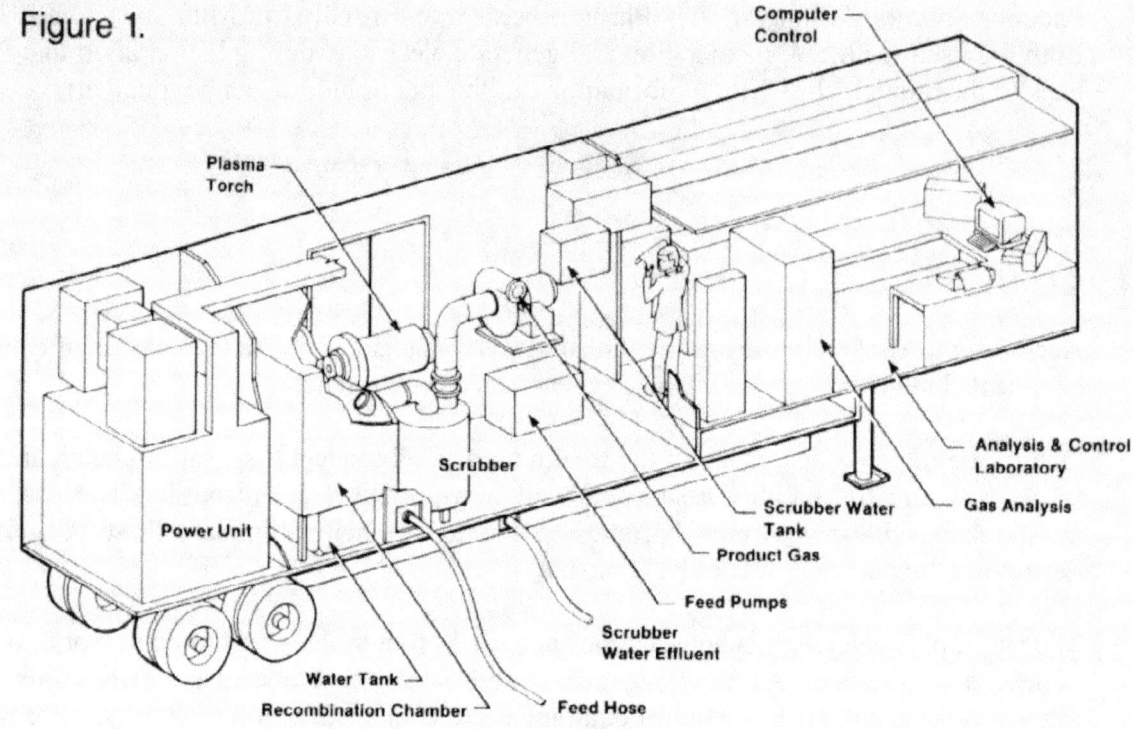

Figure 1.

Computer Control

Plasma Torch

Scrubber

Power Unit

Analysis & Control Laboratory

Gas Analysis

Scrubber Water Tank

Product Gas

Feed Pumps

Scrubber Water Effluent

Water Tank

Recombination Chamber

Feed Hose

Figure 12 – Mobile Pyroplasma Incinerator
(Drawing courtesy of Westinghouse Plasma Corp.)

The mobile unit can be used to reclaim chemical constituents from a waste stream which would normally be destroyed, stored and/or buried. The use of Pyroplasma for reclamation can reduce overall destruction and raw material costs.

Advantages of the Mobile Pyroplasma Incinerator are:

- Mobile – trailer mounted
- Demonstrated effective for wastes contaminated with PCBs
- Suitable for treatment of contaminated liquid wastes
- Destruction and removal efficiencies > 99.99999%
- Product gases usually carbon monoxide, CO and hydrogen
- H2 can be reused in the process

More on the Westinghouse Pyroplasma Incinerator can be found on their website.

PEAT International is a waste-to-resources company specializing in a proprietary **P**lasma **T**hermal **D**estruction **R**ecovery (PTDR) technology for treatment of hazardous, industrial and municipal waste. Peat has several installations in Taiwan and one in Lorton Virginia.

PEAT designed, built and proved effective a 350 kb/hr (750 lb/hr) demonstration system for the U.S. Army under the Congress funded PEPS (Plasma Energy Pyrolysis System) project in Lorton, Virginia. They successfully demonstrated the 6 to 8 ton/day capacity of the system and achieved a destruction and removal efficiency (DRE) of 99.99999%. This is well within full compliance with State and U.S. EPA requirements. Following acceptance, it was demonstrated for Agricultural Blast Media (ABM) and Regulated Medical Waste (RMW).

Upon completion of the tests, the system was dismantled and relocated to a site owned by the Georgia Tech Research Institute (a partner of PEAT), where it is used to further research and development.

PEAT also specializes in the destruction of medical wastes. For additional information on PEAT visit their website at:

http://www.peat.com

Pyrogenesis designs and builds Plasma systems for waste destruction. They have Plasma systems in use aboard U.S. Navy ships and a Carnival Cruise Lines ship.

http://www.pyrogenesis.com

LOW LEVEL RADIOACTIVE WASTE VITRIFICATION

The Westinghouse Plasma Melter provides a high-throughput, readily scalable technology for converting hazardous solid and liquid feeds into stable, low-leachability glass. Waste is screened to remove large particles with minimal contamination, partially dewatered to improve process economics, and then vitrified with the Plasma Melter to destroy the hazardous organics and convert the contaminated fines to a low-leachability glass product. The other streams leaving the system are non-hazardous discharge water, a small calcium sulfate stream from offgas sulfur removal, and a clean offgas stream.

The Westinghouse Plasma Melter has been used for processing of contaminated soil containing PCBs and feedstocks ranging from wood and construction debris to wet clay. In all cases, it produced a stable vitreous (glassy) product which passed EPA TCLP leach testing as a non-hazardous material.

In 1992 and 1993, the Plasma Melter was successfully used to demonstrate calcinations of simulated high-level radioactive waste (HLW) focusing on destruction of hazardous organics and potentially explosive ferrocyanides in complex sodium salt mixtures currently in storage by the United States Department of Energy (DOE) at the Hanford Reservation is eastern Washington State.

The success of the HLW Plasma Calcination testing lead to continued interest within the DOE and Hanford and again Westinghouse Plasma won a competitive contract to demonstrate direct conversion of tank waste stimulant to a vitrified glass product.

For more information on these tests visit the Westinghouse Plasma website.

GASIFICATION

During the last several decades, Westinghouse Plasma has conducted many successful experiments, designs and developments involving the gasification of simulated MSW, ASR (auto shredder residue), fossil fuels, and industrial liquid and solid wastes in a plasma reactor. The gasification test material feed ranged from low Btu MSW (1600 kcal/kg) to medium Btu simulated auto shredder residue (4500 kcal/kg and to high Btu coal (8,000 kcal/kg).

Experiments were conducted whereby fuels were gasified to produce primarily carbon monoxide, CO and hydrogen, H2. The inorganic components of the feed were converted to molten slag which was removed as vitrified byproduct. The slag passed the EPA-mandated Toxicity Characteristic Leachate Procedure (TCLP) requirements. The plasma torch power requirement ranged from 100 kW to 250 kW per ton/hr of MSW/ASR/Coal and higher for inorganic waste treatment/vitrification.

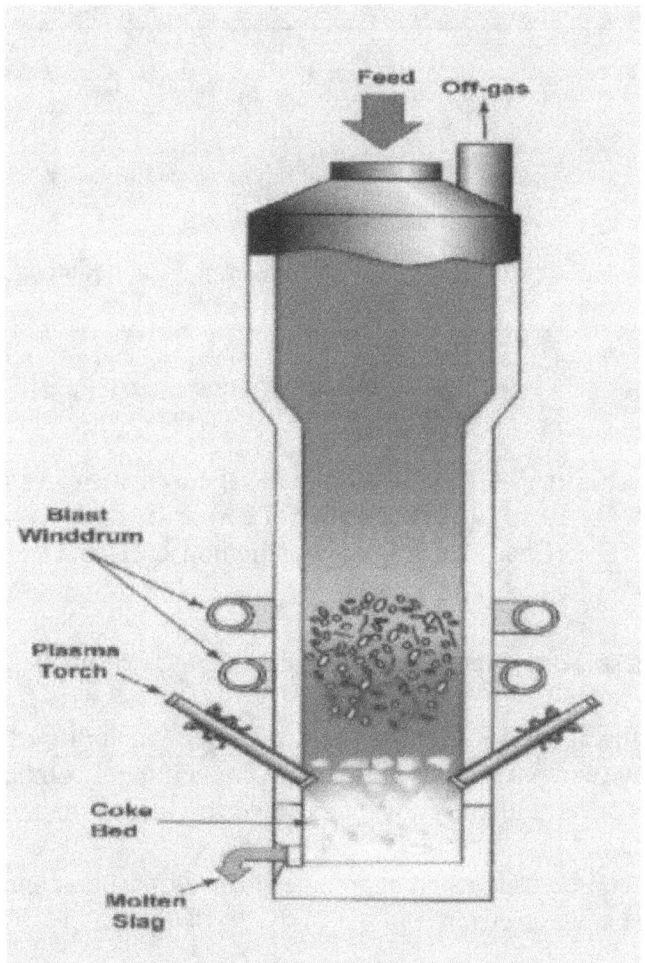

Figure 13 – Westinghouse Plasma Gasification Reactor
(Diagram courtesy of Westinghouse Plasma Corp.)

Westinghouse Plasma has designed a 300 tpd energy production and vitrification plant for another customer using a blend of MSW and industrial waste.

SUMMARY

Plasma incineration is an excellent method for destroying toxic materials including Municipal Solid Waste (MSW), Low Level Radioactive Waste and Medical Waste.

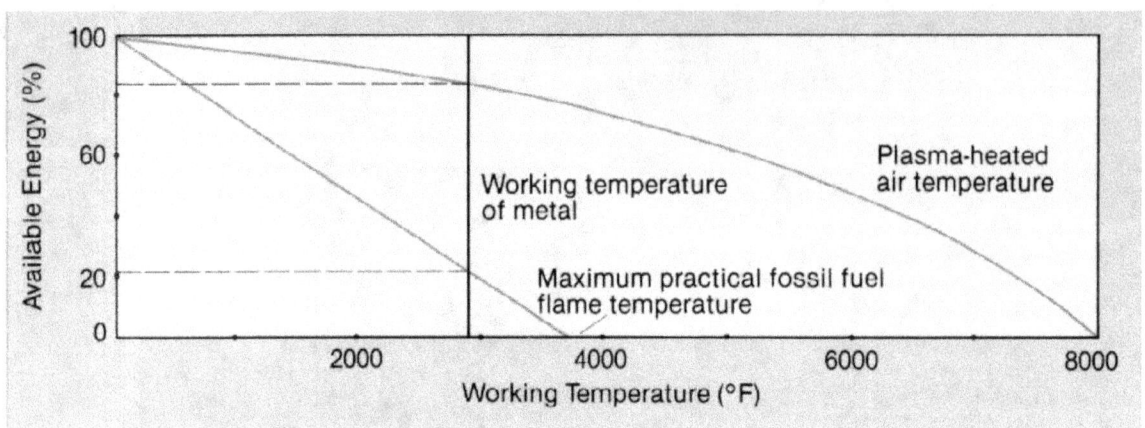

**Figure 14 – Fossil Flame Air Temperature vs Plasma
Heated Air Temperature**
(Chart courtesy of Westinghouse Plasma Corp.)

The Plasma process works well because it can generate extremely high temperatures.

As you can see in Figure 14, Plasma has the unique capability of increasing the energy of the process gas by two to ten times higher than conventional combustion equipment. Plasma technology offers several distinct advantages over traditional methods.

The by-products can be used as an aggregate for use in asphalt and other uses and the gases can be used to generate electricity.

34

Notes Chapter 4

Chapter 5

PLASMA SPRAYING

INTRODUCTION

Plasma spraying is a thermal spray process in which a non-transferred arc is the source of heat that ionizes a gas that melts the coating material and propels it to the workpiece. Plasma spraying has been around for over one hundred years. It took Dr. Max Ulrich Schoop of Zurich to recognize the potential of a stream of molten particles could create a coating. His work, along with collaborators, resulted in the thermal spray process. The thermal spray process has created a worldwide industry serving over thirty technology sectors and generating sales of over two billion dollars peryear. Dr. Schoop developed the first instrument for thermal spraying. He originally called it metallizing because he was spraying a metal onto another surface.

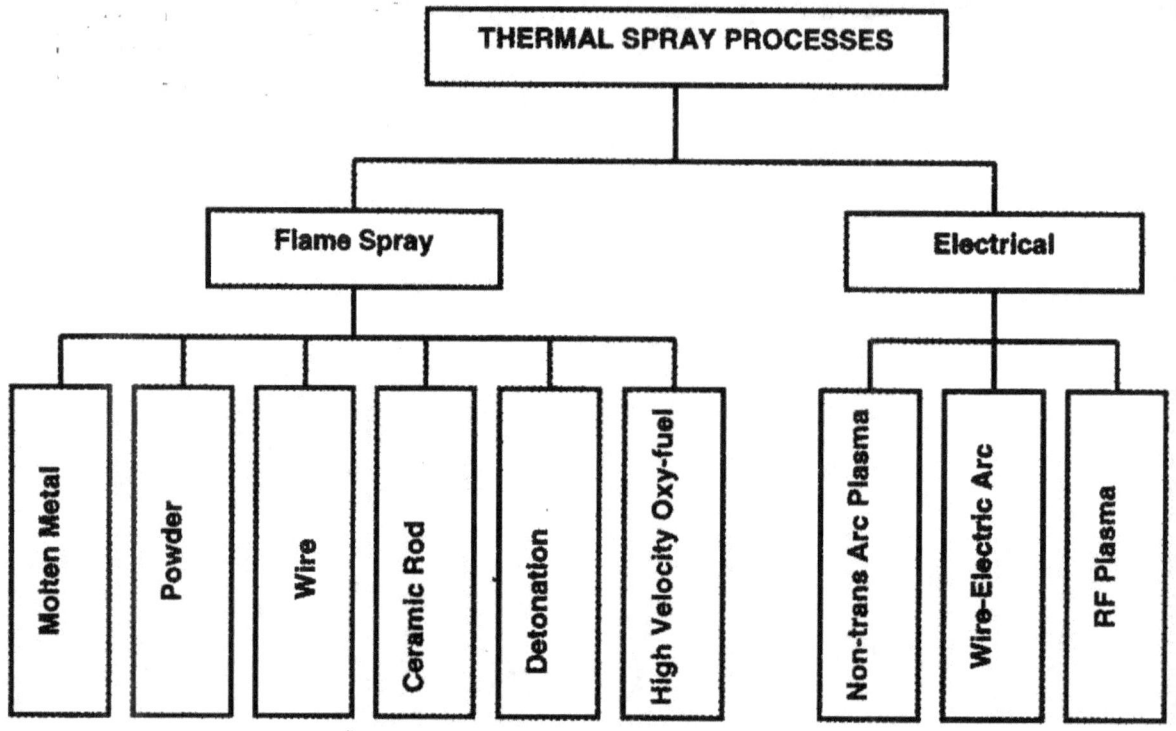

Figure 15 – Thermal spray application methods by heat source.
(Diagram courtesy of the International Thermal Spray Association "ITSA")

As you can see in Figure 15, thermal spray covers a wide variety of applications and materials. This book covers only the electrical applications.

SPRAY COATING

The coating material is usually in powder form and can be copper, aluminum, gold, ceramic materials, bronze or any other material that melts and is desirable for coating. The powder is pushed through a plasma torch using air or some other inert gas.

As the powder is passed through the electrodes of the plasma torch, it is turned into a molten material which is then sprayed on another substance such as glass. The more even the spray coating, the more uniform the color, just as if you were spray painting.

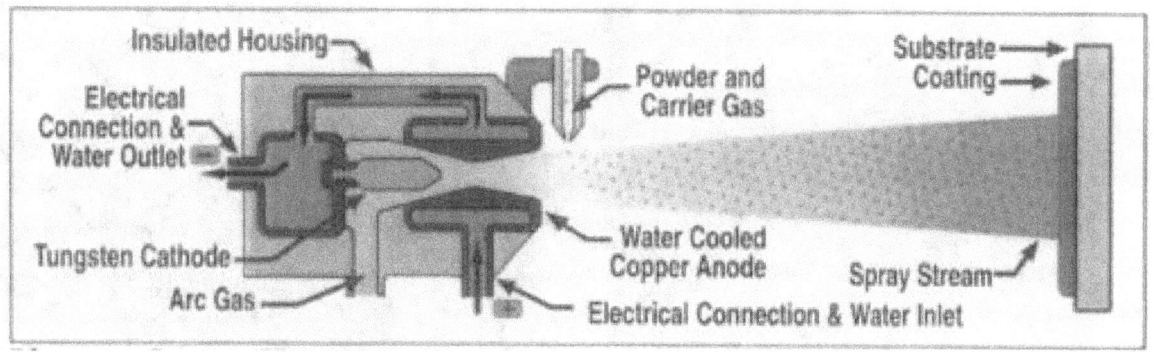

Figure 16 – Cutaway of a Plasma Spray Gun
(Diagram courtesy of the ITSA)

In the case of architectural glass, think of a high rising skyscraper. The colored glass (bronze, aluminum, etc) is most likely the result of plasma spray coating. Although other methods for coating glass are used, plasma spraying is the most common. An example of non-plasma would be thin films that are applied to the glass.

Another use for plasma spray coating is coating machined parts such as crankshafts and jet engine turbine tips. Refer to Figure 10 for an example of turbine tip blades. Advances in this area of plasma power supply design have made a real difference to the user.

Modern day plasma spray guns can produce temperatures from 5000 C (9,032 F) to 16,000 C (28,832 F) for long periods.

These are usually referred to as "non-transferred plasma generators". Typical gasses used for plasma consist of argon, nitrogen, hydrogen and helium. They can be used alone or in combinations. The gas is used to transport the powder through the plasma stream.

Figure 17 – Hand-Held Plasma Spray Coating Gun
(Photo courtesy of TAFA/Praxair)

As you can see in Figure 17, the hand-held plasma spray gun is not very large. It is shown here coating a cylindrical piece. Since the gun is small, it is easy to use for small and/or large coating applications. The guns can also be installed on a robotic arm and automatically controlled to cover large areas such as sheet glass.

The International Thermal Spray Association (ITSA) is located at:

> International Thermal Spray Association
> 208 Third Street
> Fairport Harbor, Ohio 44077

> www.thermalspray.org

The association publishes a quarterly newsletter titled "Spraytime" which includes many useful articles. Much of the information on thermal spray in this book was compiled from information provided by the ITSA.

SUMMARY

The process of thermal spraying including plasma arc spraying was initially developed in the early 1900's. The concept has grown to a sizable industry and includes many types of thermal heating. The plasma arc can melt different metals and deposit them onto another surface in a manner that is very durable, efficient, cost effective and fast.

Chapter 6

PLASMA CUTTING

INTRODUCTION

Plasma Arc Cutting has been around for a long time – since the early 1950's. It was discovered that by squeezing the arc into a very tight or small cross section, the resistance heating value of the arc was greatly increased. This process also increases the arc temperature. After passing through the plasma gun's nozzle, the arc exits as an intensely hot plasma jet stream. The addition of an inert gas such as argon increases the temperature even more. Changing the gas, the flow rate, arc current, and distance from the workpiece, etc. will change the characteristics of the plasma arc and the cut.

Figure 18 is a good example of a plasma arc cutting through a very thick piece of metal in a very short time with great accuracy. The plasma cutter works on the same principle as the plasma sprayer (described in the previous chapter) except no material is passed through the electrodes. The arc exits the plasma gun and attaches to a work piece (the piece of metal to be cut). The intense heat produced by the arc cuts through the metal in short order. Steel thickness of more than 6 inches can be cut easily with the plasma gun.

Figure 18 – Example of Plasma Cutting

CUTTING TORCHES or GUNS

All plasma torches are similar in the basic components. All have nozzles, electrodes, water chambers, gas passages, etc. as demonstrated in both Figure 1 shown in Chapter One and Figure 19 below. The mechanical design simply changes based upon the application and power requirements.

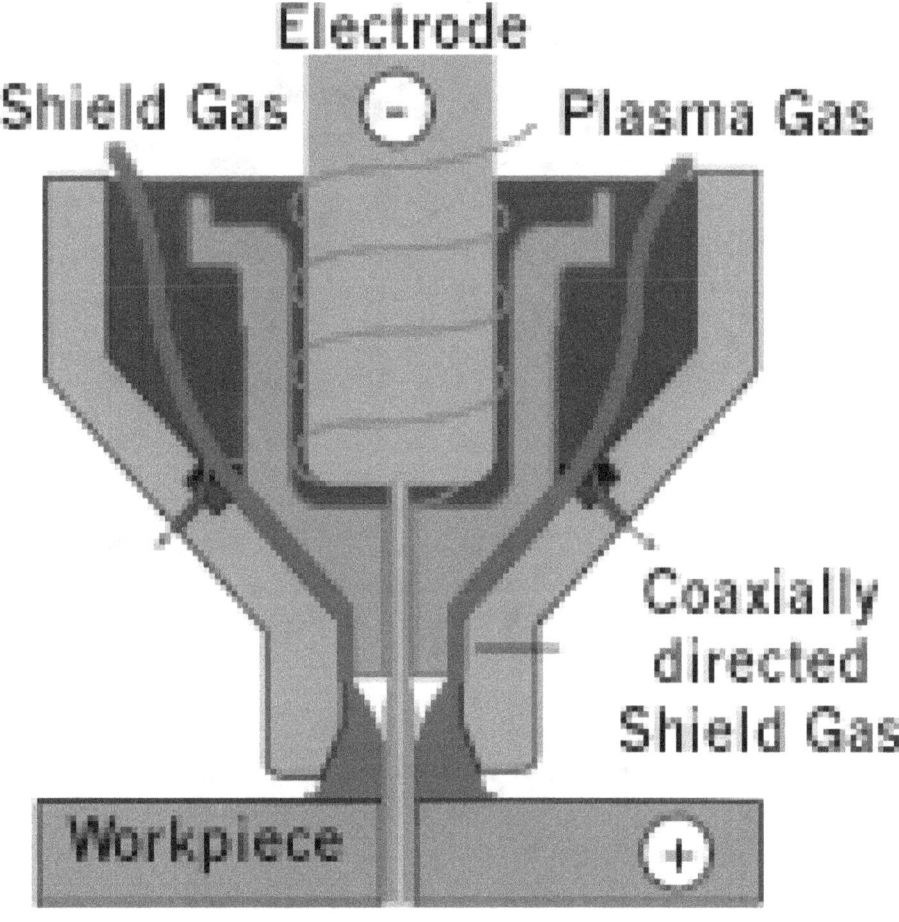

Figure 19 – Cutaway of a Plasma Cutting Torch
(Diagram courtesy of "The Fabricator.com")

Plasma cutting guns can resemble an actual garden sprayer or a gun as seen in Figure 20.

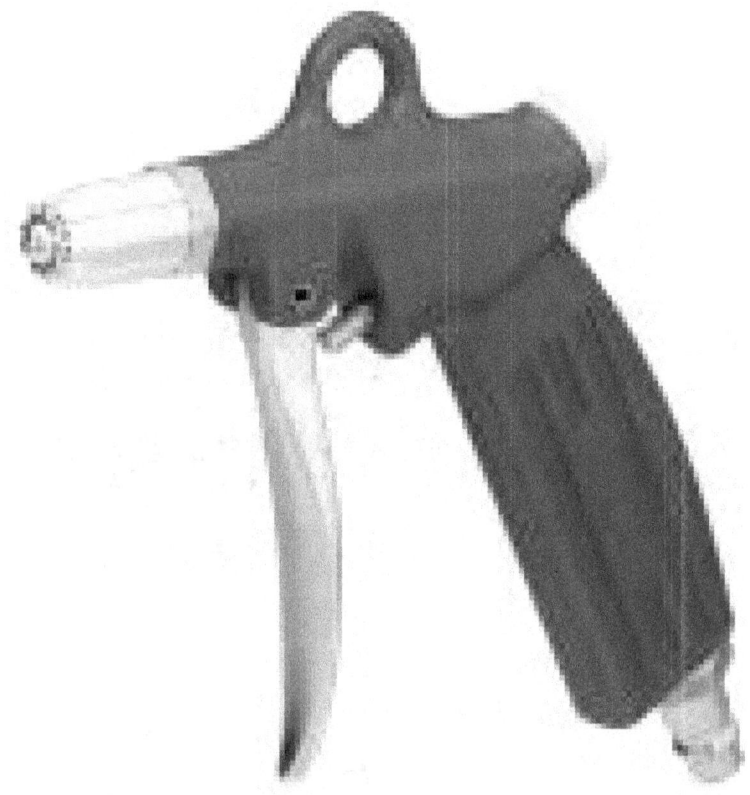

Figure 20 – Plasma Cutting Gun
(Photo courtesy of Elektra Beckum)

Plasma cutting guns can also be automated using robots as seen in Figure 21.

Figure 21 – Robot arms with Plasma Guns
(Photo courtesy of Industrial Robotix)

Since the gun operates at very high temperatures up to 20,000 degrees and gas velocity similar to a jet engine, the design of the actual plasma gun can be quite involved. The engineers and manufacturers oftentimes will go through many iterations to the design, fine tuning each of the components. The ultimate goal is a gun that is light-weight, easy to hold & maneuver and reliable with long electrode life.

SAFETY

Safety is always a concern. With the bright light, ultraviolet rays, loud noise, high electrical power and high velocity jet spray, operators can be hurt easily. Extreme care must always be taken when around a plasma torch, whether you are incinerating, spray coating or cutting. Carelessness can lead to blindness, burns and fires. Refer to the chapter on Safety and Maintenance for more details.

SUMMARY

Although there are some inherent safety issues, plasma cutting can be useful and beneficial for cutting heavy or thick metals easily. The cutting process can be either manually controlled or automated using robots.

NOTES CHAPTER 6

Chapter 7

PLASMA IN THE FUTURE

PLASMA IN THE FUTURE

Finally after more than 100 years, plasma is finally being used in more and more applications. Presently, in addition to the industrial applications already talked about in this book, we are starting to see more consumer applications. These include applications such as the plasma globes currently being sold at novelty stores to plasma television and dental/medical work.

PLASMA GLOBES

Often seen in horror movies, plasma globes have lightning (arcing) from an interior electrode to the inside of the clear globe. In some cases, the lightning can be "timed" to music, thereby giving a special effect when listening to music with the plasma globe energized.

PLASMA INCINERATION

The time is finally here. After more than a decade of Plasma Incineration installations around the world, installations in the United States are being planned. Presently, there are two commercial installation planned for the state of Florida. One in St. Lucie on the Atlantic Coast and another close to Tallahassee in the Florida panhandle. In both cases, the plants will incinerate both existing and newly created municipal waste. The energy created during the incineration will be used to create electricity. The electricity will power the plant and the excess will be sold to the local utility. The municipal waste will be reduced to appx 10% of it's original volume and the remainder is safe and can be used as an aggregate for use in asphalt roads and many other uses.

PLASMA TELEVISION

Plasma TVs are currently considered state-of-the-art. The pictures are brighter and crisper than traditional TVs and the picture almost appears three dimensional.

The plasma screen is comprised of two parallel sheets of glass. In between the sheets there is a mixture of neon and xenon gas which fills thousands of grid-like chambers.

When electricity hits the gas in the chamber, a tiny discharge of ultraviolet light is released and strikes a phosphor coating creating plasma called a pixel. The pixels create red, green and blue light (RGB). Hundreds of thousands of pixels combine to create the image on the screen. Because each pixel is so small, fantastic resolution is the result.

DENTAL AND MEDICAL

Many of us hate the thought of the Dentist drilling in our mouth. Even more so, the thought of a needle with a flaming plasma tip doesn't sound much better. However, the plasma needle is cold and painless to the touch. The needle's creator, physicist Eva Stoffels-Adamowics who is based at the Eindhoven University of Technology in the Netherlands, says it could also be used to painlessly remove cancerous tissue. Dental treatment of the future? You be the judge.

CONCLUSION

We are seeing more and more applications for home such as the plasma globe and plasma TV. The real good news is plasma incineration. Not only does plasma clean-up landfills, it incinerates newly created "trash" safely, creates it's own energy source with spare electricity for other uses. One use is to feed it back into the local electric utility grid. Since it generates it's own electrical power, I think of it much like perpetual motion. As long as you feed it trash, it requires no outside energy source. The remaining materials can be used safely for other applications rather than being left in landfills, polluting our environment.

Plasma Incineration could solve many of the environmental issues we are faced with today. Plasma is clearly a choice to be considered now, not tomorrow!

NOTES CHAPTER 7

SECTION TWO

PLASMA POWER SUPPLIES

Chapter 8

TYPES OF PLASMA POWER SUPPLIES

INTRODUCTION

This chapter is only intended to be an introduction to the different types of Plasma Power Supplies. Later chapters will describe in detail each of the most significant types of power supplies for use in industrial plasma applications. Each type has its own operating characteristics, advantages and disadvantages. Many engineers over the years have applied "off the shelf" power supplies to plasma applications with minimal success. Their lack of success is attributable to the dynamic resistance characteristics of a plasma arc.

The majority of DC power supplies is designed to operate into static loads and simply will not perform like a custom engineered plasma power supply will.

Although both AC and DC output power supplies can be used for industrial plasma applications, only the DC type will be discussed due to the better performance and popularity of DC control.

Plasma Power Supplies typically operate in the current control mode although I have designed and built power regulated units. The use of current regulation maintains a constant current regardless of the arc resistance changing due to temperature and/or gas volume changes. It also compensates for incoming power fluctuations.

WELDING POWER SUPPLY

The welding power supply is the simplest of the different power supplies used for plasma applications. When used in the welding industry, welding power supplies are typically moved from location to location. They are not faced with the same obstacles plasma heating applications face. These obstacles include 100% duty cycle, high power factor requirements, low DC ripple, precise DC current regulation, remote control and reduced power harmonics.

The welding power supply consists of the following main components:

- Input Power Switch
- Step-down isolation transformer
- Rectifier
- DC choke
- Electronic circuits

It relies upon the magnetic circuits (transformer and choke) to set the maximum output voltage and to accomplish the DC current regulation. The electronic circuits perform a coarse output adjustment while the magnetic circuits do the work from there. This works well for welding applications but is not the best method for industrial plasma applications. It is also the simplest and least expensive type of plasma power supply.

SCR FAST REGULATOR POWER SUPPLY

The SCR fast regulator power supply consists of the same main components as the welding power supply except it relies more upon the electronic circuits for current regulation than the magnetic circuits. The electronic circuit samples the output DC current, compares it to a set-point signal and regulates the DC current very quickly. The magnetic circuits are used mainly for setting the maximum output voltage and for short circuit protection and DC ripple improvement.

This type of power supply can be used for welding as well as plasma applications. It has improved ripple & DC current regulation, higher power factor, reduced harmonics, is 100% duty cycle rated and allows control from a remote location. It is definitely a major improvement over welding power supplies. See Figure 22 for an example of a typical SCR Fast Regulator Power Supply.

PWM BUCK REGULATOR POWER SUPPLY

The PWM Buck Regulator Power Supply is an extension of the SCR Fast Regulator by adding an additional stage. It begins with the same circuit topology and inserts a DC Filter and a Buck Regulator between the rectifier and the DC Choke.

PWM stands for Pulse Width Modulator. The PWM Buck Regulator turns the DC voltage on and off (pulses) at a predetermined frequency. It also changes the time on and time off (width) proportionally to regulate the output current. This type of power supply has even better power factor, reduced harmonics, better regulation and smoother and/or more stable arc current than the SCR Fast Regulator.

Figure 22 – Typical SCR Fast Regulator Power Supply
(Photo courtesy of Hardface Alloys Inc.)

SUMMARY

Each of the three most common types of plasma power supplies has its own advantages and disadvantages. The welder is the least expensive with the lowest performance characteristics. The PWM Buck Regulator is the most expensive but has the best operating characteristics of the three. The SCR regulator is somewhere in the middle for cost and performance; however, it is the most popular. Each has its own applications in which it performs best. The following two chapters will cover the "SCR" and the "PWM" type of power supplies in detail.

NOTES CHAPTER 8

Chapter 9

SCR FAST REGULATOR PLASMA POWER SUPPLIES

GENERAL

The SCR Fast Regulator is the most popular type of Plasma Power Supply that I have designed. It is simplistic in design (similar to a Welding Supply) and yet has much better performance than a welder. It normally operates in the DC current regulation mode; however, It is possible to built plasma power supplies that operate in the voltage and power regulation mode also.

MAIN COMPONENTS

There are five main components:

- Input Switchgear
- Isolation Transformer
- Rectifier
- DC Choke
- Electronic Circuits

INPUT SWITCHGEAR - This component provides isolation from the power source. It can be a disconnect switch, contactor or circuit breaker. The disconnect switch can be fused or un-fused and allows the user to manually switch the power source on or off. The un-fused disconnect switch provides no short circuit protection unless it is fused.

The contactor allows the user to switch the incoming power on or off via locally mounted pushbuttons, a switch or via a

Figure 23 – Thermal-Magnetic Circuit Breaker
(Photo courtesy of Cutler-Hammer)

remote signal from a computer for example. Like the un-fused disconnect switch, the contactor does not provide any short circuit protection.

The circuit breaker is much like a disconnect switch except it provides instantaneous short circuit and steady-state over current protection. Most circuit breakers are thermal-magnetic. This means it will open the circuit in the case of a steady state over-current situation (thermal) or a instantaneous short circuit (magnetic). The circuit breaker is operated locally via a mechanical handle or a remote shut-down signal. Most circuit breaker manufacturers offer motor operators for remote on/off control. However, motor operators tend to be expensive and are not very popular.

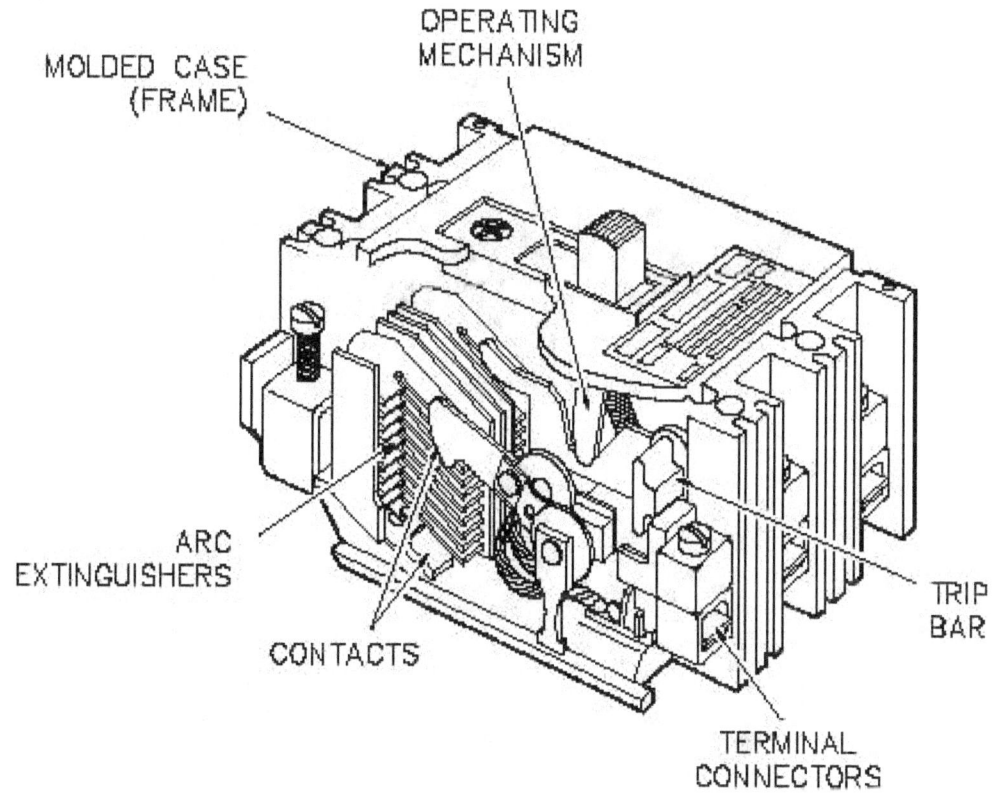

Figure 24 – Cutaway of a thermal-magnetic Circuit Breaker
(Diagram courtesy of Integrated Publishing)

65

Some users combine the circuit breaker and the contactor. In this case, the circuit breaker provides the protection and the contactor allows for remote on/off operation.

If the power supply has input power over 600 Vac, a medium voltage disconnect switch (fused)contactor combination is generally used. In the case of input voltages in excess of 7.2kV, special switchgear is used.

ISOLATION TRANSFORMER - Most all plasma torches are referenced to earth ground (grounded) and nearly all power sources are also grounded. This creates a requirement for isolation. The power supply manufacturer should provide an isolation transformer to eliminate the double grounding and to convert the incoming voltage level to that required by the

Figure 25 – Three-Phase Isolation Transformer
(Photo courtesy of Control Transformer Co.)

rectifier, and ultimately, the torch. For example, the incoming power may be 480 Vac but only 300 Vac is required by the rectifier. The isolation transformer steps down the voltage. Of course, the opposite could also be true. If a higher voltage is required by the rectifier than is available from the power source, the transformer steps up the voltage.

Without getting too technical, the transformer also provides multiple outputs. Multiple outputs are required when minimizing the harmonics (see chapter on Harmonics) is important or if reducing the output ripple is a requirement of the plasma system.

RECTIFIER - The rectifier is the electrical portion of the power supply that converts the AC power into DC power. It also allows the user to adjust the output power level of the power supply. In the SCR fast regulator power supply, the conversion is accomplished using SCRs (Silicon Controlled Rectifier). For a more thorough understanding of SCRs, refer to my book An Introduction to SCR Power Controls available at:

www.amazon.com

or

http://www.cafepress.com/powercontroller

If the power supply has three-phase incoming power (most do), the rectifier is made using multiples of six SCRs. Each SCR provides one pulse of AC to DC conversion. The more sets of six SCRs, the less harmonics and ripple are produced. Simple or less expensive plasma power supplies use only six pulse conversion. More complicated power supplies use 12, 18, 24 or more pulses to help minimize harmonics and ripple.

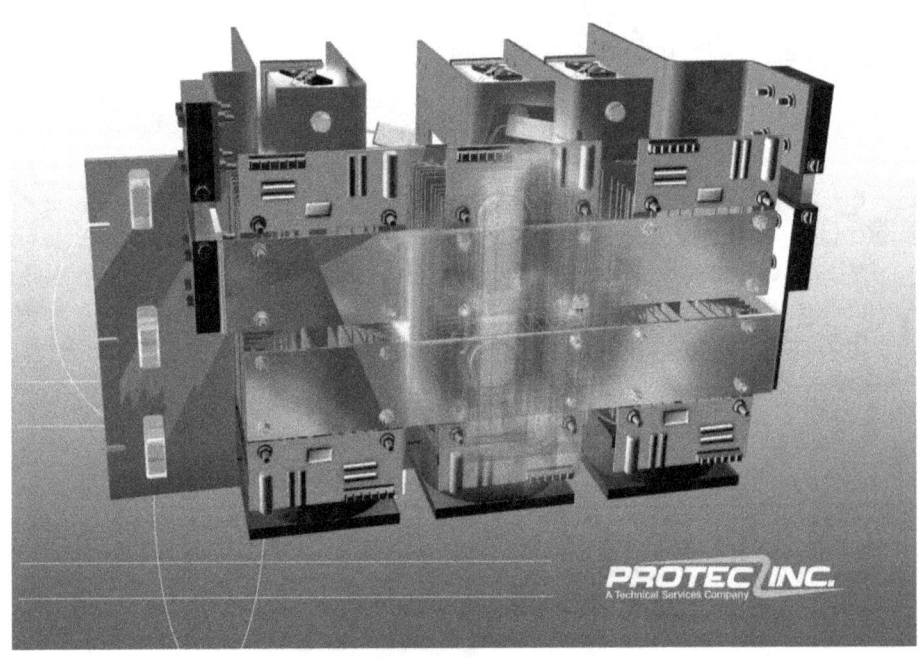

Figure 26 – Typical 6-Pulse Rectifier
(Drawing courtesy of Protec, Inc.)

DC CHOKE - The DC choke performs three primary functions in a plasma power supply.

1. Helps limit short circuit currents
2. Helps minimize ripple in the DC current
3. Helps minimize over-shoot on start-up

The DC choke is a magnetic device much like the transformer. Basically, the choke fights any change in DC current and that limits the short circuit currents and the over-shoot on start-up. It also acts as a filter and helps to smooth the DC current. Sizing of the choke is critical to the plasma power supply's proper operation. A choke too big or too small can cause dramatic differences in the plasma torch operation. The choke's inductance rating is based upon the DC voltage, ripple and DC current.

ELECTRONIC CIRCUITS - The electronic circuits are the brains of the plasma power supply. These circuits sample the incoming AC power, the output DC power and command signals from a user and adjust the output current to that required by the plasma process. Welding power supplies control the plasma arc using the electronics to perform a course adjustment of the output. The magnetics do much of the DC current regulation.

The fast regulator method that I helped develop in the mid 1970's clearly does a better job of stabilizing the DC current. Stabile DC current is critical to plasma processes and to the life of the plasma torch.

As you can see in the oscilloscope tracing in Figures 27 and 28, the fast regulator approach (Figure 28) provides a much smoother DC current.

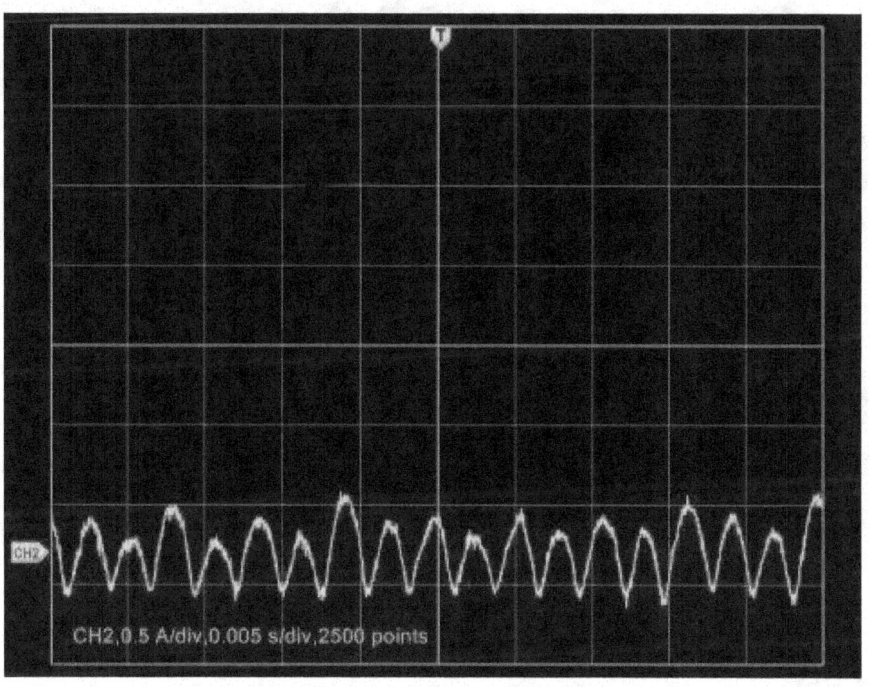

F0000CH2.CSV

Figure 27 – DC Current using a traditional Welder
(Photo courtesy of HDR Power Systems)

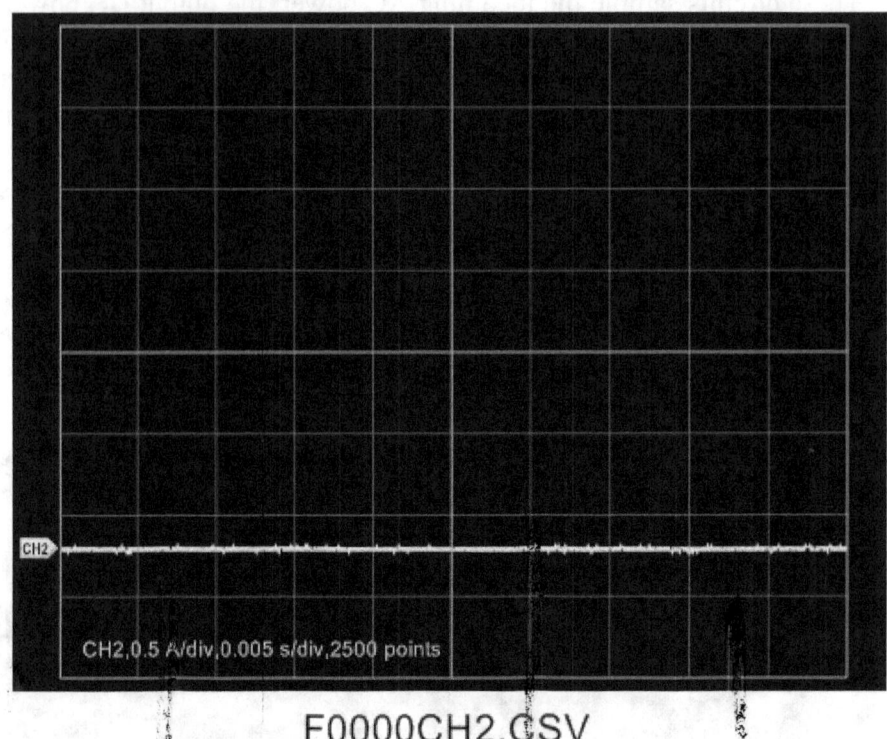

F0000CH2.CSV

Figure 28 – DC Current using a fast regulator
(Photo courtesy of HDR Power Systems)

Reading the fine print in each of the oscilloscope tracings indicates the scope measured the DC current with the calibration settings the same.

The advantages and disadvantages are described in the following chapter.

SUMMARY

Different types of SCR controlled plasma power supplies can produce different results in the plasma process. Clearly, the SCR fast regulator regulates the DC current better than a welder power supply.

NOTES CHAPTER 9

Chapter 10

PWM BUCK REGULATOR
PLASMA POWER SUPPLIES

GENERAL

The PWM buck regulator plasma power supply is an extension of the SCR fast regulator. It is basically an SCR power supply with two major components added, a DC filter and a PWM buck regulator. plus associated control circuitry. Refer to Figure 29 to see the schematic differences between the two types of power supplies. The DC filter and buck connected between the rectifier and the DC choke.

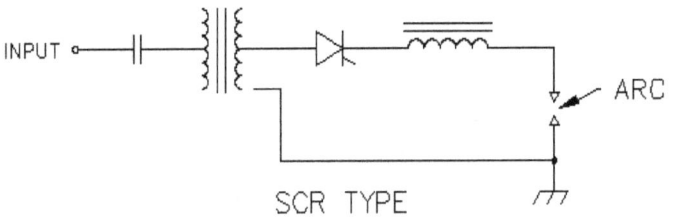

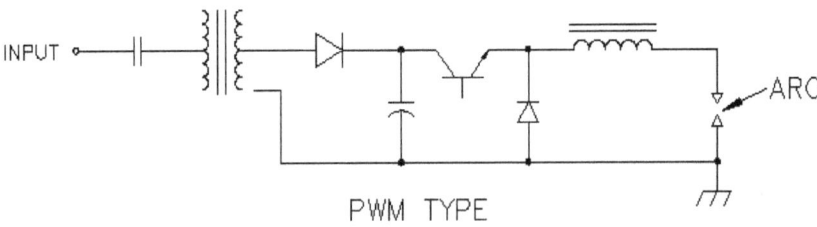

Figure 29 – One-line Schematic of SCR Fast Regulator (top) and a PWM Buck Regulator Plasma Power Supply (bottom)

In the SCR fast regulator, the SCRs control the plasma arc by increasing and decreasing the power supply's output as required by the process. In the PWM supply, the SCRs simply ramp up the voltage to charge the DC filter and then remains at full output voltage. The buck regulator performs the control by increasing and decreasing the output current as required.

ADDITIONAL MAIN COMPONENTS

DC FILTER - The DC filter provides smoothing of the DC current out of the rectifier. The more pulses the rectifier has, the less filtering is required. Increasing the DC filtering reduces the DC ripple subjected to the PWM buck regulator. This is very desirable because it helps increase the life of the PWM switching device.

Figure 30 – Typical PWM Buck Regulator Power Supply
(Photo courtesy of HDR Power Systems)

PWM BUCK REGULATOR - The initials "PWM" stand for "Pulse-Width-Modulated". This is a method of turning the power supply's output on and off in a very controlled manner quickly. The turning on and off is the "Pulse". Varying the amount of time on vs time off is the "Width". The combination is Pulse-Width-Modulation.

ADVANTAGES/DISADVANTAGES

The advantages of PWM vs SCR control are determined by the application. If the application requires low harmonic generation, reduced RFI (radio frequency interference), high power factor, reduced ripple or any one of these, then you likely need a PWM Buck Regulator. If not, then an SCR fast regulator will do the job for you.

Advantages/Disadvantages of SCR/PWM DC Power Supplies		
Advantage/Disadvantage	SCR	PWM
Output Ripple	higher	lower
Power-Factor	lower	higher
kVA Requirements	higher	lower
Open Circuit Voltage	higher	lower
Circuit Complexity	less	more
Harmonics	higher	lower
RFI	higher	less
Size	smaller	slightly larger
Cost	lower	higher

Figure 31 – Advantages/Disadvantages Chart

Portable plasma cutting equipment is a good example of an application for an SCR Power Supply. Coating of jet engine turbine tip blades is an excellent example of an application using a PWM Plasma Power Supply. These applications will provide great results. See Figure 31 for a comparison between the two types of plasma power supplies.

SUMMARY

PWM buck regulator power supplies offer several advantages over the SCR fast regulator. However, the application and initial cost allowances usually determine the type of power supply used.

NOTES CHAPTER 10

Chapter 11

IEE-519, POWER FACTOR, HARMONICS & OPERATING COSTS

INTRODUCTION

The bold print is the foreword as taken from IEEE Standard 519-1992 "IEEE Recommended Practices and Requirements for Harmonic Control in Electrical Power Systems".

This recommended practice was prepared by a joint task force sponsored by a joint task force sponsored by the Working Group on Power System Harmonics of the Transmission and Distribution Committee of the IEEE Power Engineering Society and

the Harmonic and Reactive Compensation Subcommittee of the Industrial Power Conversion Committee of the IEEE Industry Applications Society. This recommended practice is an update of the IEEE guide that was published in 1981. The work to revise the guide was started in 1984 and has incorporated the evolving understanding of the effect of static power converters and other nonlinear loads on electric power systems.

This recommended practice recognizes the responsibility that users have not to degrade the voltage of the utility serving other users by requiring nonlinear currents from the utility. It also recognizes the responsibility of the utilities to provide users with close to a sine wave of voltage. The recommended practice suggests guidelines for accomplishing this.

The recommended practices of IEEE-519 intend to set goals for electrical systems that include both linear and nonlinear loads. It describes the voltage and current waveforms that may exist and also sets goals for waveform distortion at the point of common coupling.

Nonlinear loads change the sine wave shape of the AC current and increase the AC voltage drop. This, in turn, creates or increases the harmonic currents in the AC power system.

TYPES OF NONLINEAR LOADS

Nonlinear load power supplies are important because they can be used to convert AC to DC, DC to AC, DC to DC and AC to AC. There are many types of nonlinear loads including but not limited to the following:

- Electrochemical Rectifiers
- Uninterruptible Power Supplies (UPS)
- Adjustable Frequency Drives for motors
- Arc or Plasma Power Supplies
- Saturating Magnetic Transfomers such as Saturable Core Reactors and Variable Reactance Transformers (VRT)

Electrochemical Rectifiers – These rectifiers are usually electrically very large. DC currents as high as 40 or 50 thousand amps are common. DC voltages are typically below 500Vdc. Circuit configurations are typically 6 or 12 pulse using interphase transformers to help balance the AC currents. Typical applications include plating and chlorine production.

Uninterruptible Power Supplies (UPS) – These power supplies are used widely in both Data Processing and industrial applications. A typical UPS consists of a rectifier, DC filter, inverter, battery charger, bank of batteries and usually an isolation transformer. A typical application is supplying steady and reliable AC power to computers during unwanted/unexpected power outages and/or emergency shutdowns.

Adjustable Frequency Drives - These consist of a rectifier, DC filter, variable frequency inverter and sometimes an isolation transformer. By varying the frequency of the AC synchronous motor's incoming power, the speed will vary in relationship to that frequency. These are manly used for industrial processes where precise speed control is required such as conveyors. They are commonly referred to as a "VFD" for Variable Frequency Drive.

Arc or Plasma Power Supplies – These power supplies consist of incoming power isolation transformers, SCR controlled rectifiers (6, 12, 24 pulse) and a current stabilization reactor. A pulse width modulated (PWM) power supply will include some method to modulate the DC output. These are typically used by steel mills, plasma spray coating, incineration and any other application where controlled extremely high temperatures are required such as metals research.

Saturable Core Reactors and Variable Reactance Transformers – These are magnetic devices that allow the user to control various amounts of electrical power using a very small electrical control device such as an SCR Power Control. Both are similar in concept and are typically used in applications requiring high amounts of controlled power that are likely to have shorting in the load. Saturable Core Reactors and Variable Reactance Transformers are self current limiting.

POWER FACTOR

Power Factor is basically a utilization factor. In very simple terms, power factor is the ratio of the kW divided by the kVA. A power factor of unity or 1, is perfect. An example of a poor power factor would be a process using 200 volts out of an available 400 volts. The power factor in this case would be expressed as 0.5 which is very low. This means you are using only one-half of the power you are paying for.

HARMONICS

Harmonics are associated with the switching action of solidstate control devices such as SCRs. When the SCR is phased back to a low conduction angle, high harmonics are generated. The SCR fast regulator and PWM plasma power supplies produce less harmonics than a typical welding power supply due to the higher operating voltage as a percent of the open circuit voltage. By increasing the number of pulses in a rectifier, the harmonics are reduced. Harmonic currents can cause additional heating in transformers and power cables and eventually cause early equipment failure.

OPERATING COSTS

Some operator/owners consider only the initial cost as the total cost of ownership for a particular piece of equipment. However, operating costs should be considered along with the initial cost of purchase and any money saved during the initial purchase.

The difference in operating cost for the different types of plasma power supplies can be quite significant. The SCR fast regulator operates at a higher percentage of the open circuit voltage which means the input power requirements are lower and the power factor (P.F.) is higher. You can compare the estimated yearly operating cost of a traditional welding power supply and a SCR fast regulator by referring to the charts in Figures 32 and 33. The charts assume operation at approximately 80 kW.

Hrs. operation / year	750	1000	1500	2000	2500	3000
KWH Cost						
0.02	$2,550	$3,400	$5,100	$6,800	$8,500	$10,200
0.03	$3,825	$5,100	$7,650	$10,200	$12,750	$15,300
0.04	$5,100	$6,800	$10,200	$13,600	$17,000	$20,400
0.05	$6,375	$8,500	$12,750	$17,000	$21,250	$25,500
0.06	$7,650	$10,200	$15,300	$20,400	$25,500	$30,600
0.07	$8,925	$11,900	$17,850	$23,800	$29,750	$35,700
0.08	$10,200	$13,600	$20,400	$27,200	$34,000	$40,800
0.09	$11,475	$15,300	$22,950	$30,600	$38,250	$45,900
0.10	$12,750	$17,000	$25,500	$34,000	$42,500	$51,000
0.11	$14,025	$18,700	$28,050	$37,400	$46,750	$56,100
0.12	$15,300	$20,400	$30,600	$40,800	$51,000	$61,200
0.13	$16,575	$22,100	$33,150	$44,200	$55,250	$66,300
0.14	$17,850	$23,800	$35,700	$47,600	$59,500	$71,400
0.15	$19,125	$25,500	$38,250	$51,000	$63,750	$76,500
0.16	$20,400	$27,200	$40,800	$54,400	$68,000	$81,600
0.17	$21,675	$28,900	$43,350	$57,800	$72,250	$86,700
0.18	$22,950	$30,600	$45,900	$61,200	$76,500	$91,800
0.19	$24,225	$32,300	$48,450	$64,600	$80,750	$96,900
0.20	$25,500	$34,000	$51,000	$68,000	$85,000	$102,000

Figure 32 – Estimated Yearly Operating Cost for a Traditional Welding Power Supply

Hrs. operation / year	750	1000	1500	2000	2500	3000
KWH Cost						
0.02	$1,710	$2,280	$3,420	$4,560	$5,700	$6,840
0.03	$2,565	$3,420	$5,130	$6,840	$8,550	$10,260
0.04	$3,420	$4,560	$6,840	$9,120	$11,400	$13,680
0.05	$4,275	$5,700	$8,550	$11,400	$14,250	$17,100
0.06	$5,130	$6,840	$10,260	$13,680	$17,100	$20,520
0.07	$5,985	$7,980	$11,970	$15,960	$19,950	$23,940
0.08	$6,840	$9,120	$13,680	$18,240	$22,800	$27,360
0.09	$7,695	$10,260	$15,390	$20,520	$25,650	$30,780
0.10	$8,550	$11,400	$17,100	$22,800	$28,500	$34,200
0.11	$9,405	$12,540	$18,810	$25,080	$31,350	$37,620
0.12	$10,260	$13,680	$20,520	$27,360	$34,200	$41,040
0.13	$11,115	$14,820	$22,230	$29,640	$37,050	$44,460
0.14	$11,970	$15,960	$23,940	$31,920	$39,900	$47,880
0.15	$12,825	$17,100	$25,650	$34,200	$42,750	$51,300
0.16	$13,680	$18,240	$27,360	$36,480	$45,600	$54,720
0.17	$14,535	$19,380	$29,070	$38,760	$48,450	$58,140
0.18	$15,390	$20,520	$30,780	$41,040	$51,300	$61,560
0.19	$16,245	$21,660	$32,490	$43,320	$54,150	$64,980
0.20	$17,100	$22,800	$34,200	$45,600	$57,000	$68,400

Figure 33 – Estimated Yearly Operating Costs for a SCR Fast Regulator Plasma Power Supply

For example; an SCR Fast Regulator power supply (whether SCR or PWM controlled) operating for 3000 hours per year using $0.10 per KWH power results in a yearly operating cost savings of $16,800 ($51,000 - $34,200). Users in areas with higher electrical charges can save even more. Of course, if the power supply operates for more than 3000 hours per year the savings is even greater.

In addition to the direct savings associated with the input power usage savings of an SCR fast regulator, the increased Power Factor reduces the operating costs even more. Some, but not all, electrical utilities add additional charges when the user's power factor is low. By using the SCR fast regulator or PWM power supply, the additional charges for low power factor can be eliminated or reduced.

Never consider just the initial cost of a plasma power supply as the total cost without looking further into the operating costs and process improvements. There are considerable savings to be realized during the operation of this equipment by choosing the proper type of power supply.

SUMMARY

Choosing an improper type of plasma power supply can create problems for the installation, such as poor Power Factor or increased harmonics. Lower Power Factor associated with typical welder power supplies can end up costing the operator considerably more than the initial cost savings for a lower cost power supply. There can be other operating cost savings above the normal operating costs by using the SCR fast regulator or PWM power supply.

If meeting the requirements of IEEE-519 is important, you must use at least 18 pulses or more of SCR control. If using PWM control, you may be able to use only 12 pulses.

NOTES CHAPTER 11

Chapter 12

STARTING THE PLASMA TORCH

GENERAL

The plasma arc occurs between two electrically conductive pieces usually called electrodes or the workpiece. The gap between these is usually air, an inert gas or vacuum in which all act as an insulator. Electricity does not flow through an insulator. A device called a starter or an arc ignitor is required to jump start the plasma arc.

STARTING THE TORCH

Once the process is ready, the power supply's output is turned on. At this point, the power supply is providing full open-circuit (VOC) voltage but no current is flowing. In order to obtain arc current, the plasma torch's gas must be broken down. Ignition occurs when the arc ignitor provides a high voltage. This high voltage breaks down the torch's gap and current flows. Some plasma applications require an inert gas to be passed through the gap to aid ignition. This can be helium, argon, nitrogen, hydrogen or one of the many other gases available. Once the plasma arc is established, the process may change from a starting gas to another gas that has a higher heat value. This is strictly based upon the application.

STARTER/ARC IGNITOR

The high voltage/high frequency ignitor is the most common type of starter in use today and is the only one covered in this book.

The starter consists of the following main components: high voltage ignition transformer, isolating coil, spark gap and two RC networks.

When energized, the high voltage ignition transformer produces a high voltage, usually 3000 to 12000 volts. The circuit consisting of this transformer, the spark gap and one of the RC networks crate a high voltage/high frequency voltage. This voltage is applied to the torch's gap via the isolating coil. This high voltage breaks down the torch gap and current flows. A second RC network that is connected across the power supply's output terminals discharges its stored energy and helps start the torch. This RC network was previously charged when the power supply was first turned on.

Once the arc is established, the user can adjust the output level to that required by the process. See Figure 34 for a typical Arc Ignitor.

Figure 34 – High Frequency/High Voltage Arc Ignitor
(Photo courtesy of Westinghouse Plasma Corp.)

CHOKE-BYPASS

Earlier in the discussion on the DC choke, recall that the choke resists changes in DC current. This can hinder starting because we are attempting to go from zero current to some predetermined current level. In the late 1970's, I developed the choke-bypass circuit.

Very simple in design, the circuit simply puts a low ohm resistor in parallel with the DC choke. During starting, the power supply's DC current chooses the bypass resistor as the path of least resistance. Once the arc current is established, the resistor is disconnected from the circuit via a bypass contactor, the arc current flows through the DC choke.

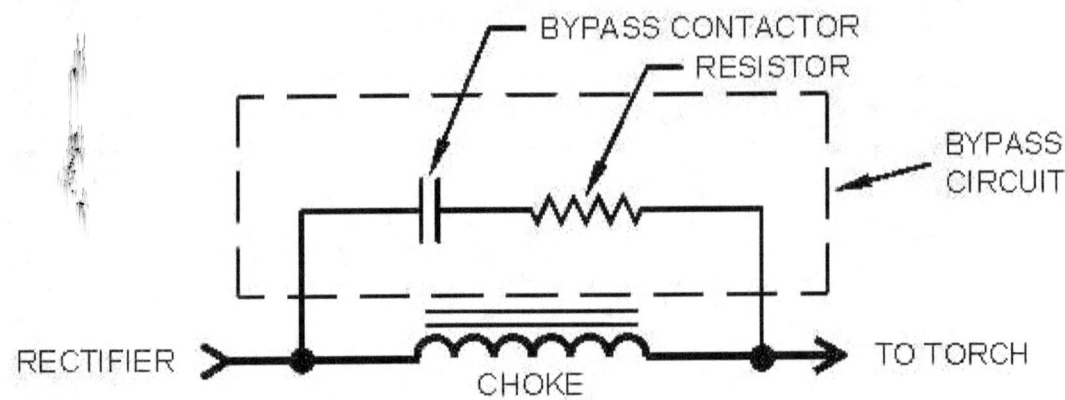

Figure 35 – Schematic of a Choke Bypass Circuit

The low power rated plasma processes (such as plasma cutting) typically do not need choke-bypass circuits and, in some cases, no ignitor. Nearly all higher powered (greater than 1 mega-watt) plasmas require one.

The choke-bypass circuit not only helps to ensure easy starting but also helps increase electrode life by minimizing over-shoot of the DC arc current during starting. See Figure 35 for the schematic of a choke bypass circuit.

SUMMARY

Although not always needed, the ignitor helps users to ignite torches reliably. Different gasses help also. A choke-bypass circuit (usually used on large plasma systems) can also help start the arc and help lengthen the life of the plasma torch.

NOTES CHAPTER 12

Chapter 13

SAFETY and MAINTENANCE

GENERAL

Unlike many types of electrical equipment, plasma equipment has safety issucs such as impacting vision or causing severe burns. Make sure any person working on the plasma equipment is trained in the proper methods for service and operation of the equipment. Failure to do so can result in equipment and/or personnel injury.

SAFETY

It is important to follow all safety guidelines and use proper safety gear at all times. Refer to Diagram 36 for the safety symbols usually associated with any type of plasma equipment.   Much of the safety information in this chapter has been provided by Hypertherm, Inc. located at Etna St. P.O. Box 5010 in Hanover, N.H. 03755 USA.

Figure 36 – Typical Warning Symbols

(Diagram courtesy of Hypertherm Corp.)

See the following page for information on each of the symbols. Even though many of these pertain to plasma cutting, they provide a good point of reference for the safety issues involved. For more complete information on safety as related to plasma cutting, contact Hypertherm, Inc. The numbered text corresponds to the numbered boxes on the label. Warning labels are not always supplied on plasma equipment so always read the instruction manual first.

1. Cutting sparks can cause explosion or fire.

1.1 Keep flammables away from cutting.

1.2 Keep a fire extinguisher nearby, and have a watchperson ready to use it.

1.3 Do not cut on any closed containers.

2. The plasma arc can cause injury and burns.

2.1 Turn off power before disassembling torch.

2.3 Do not hold the material near cutting path.

3. Electric shock from torch or wiring can kill. Protect yourself from electric shock.

3.1 Wear insulating gloves. Do no wear wet or damaged gloves.

3.2 Insulate yourself from work and ground.

3.3 Disconnect input plug or power before working on machine.

4. Breathing cutting fumes can be hazardous to your health.

4.1 Keep your head out of the fumes.

4.2 Use forced ventilation or local exhaust to remove the fumes.

4.3 Use ventilating fan to remove the fumes.

5. Arc rays can burn eyes and injure skin.

5.1 Wear hat and safety glasses. Use ear protection and button shirt collar. Use welding helmet with correct shade of filter. Wear complete body protection.

6. Become trained and read the instructions before working on the machine or cutting.

7. Do not remove or paint over (cover) the warning labels.

EYES, SKIN and HEARING DAMAGE

Plasma arcs produce intense visible and invisible (ultraviolet and infrared) rays that can burn eyes and skin. Always use eye protection in accordance with applicable national and/or local codes. Eye protection (safety glasses or goggles with side shields, or a welding helmet) with the appropriate lens shading to protect your eyes. Refer to the equipment manufacturer for the proper shading information.

Wear protective clothing to protect against burns caused by ultraviolet light. Protective clothing should include safety shoes, gloves, hat, flame retardant clothes covering all exposed skin and never carry any combustibles, such as a butane lighter or matches.

Prolonged exposure to noise associated with plasma torches can damage hearing. Always wear approved ear protection and warn others nearby about the noise hazard.

There are many standards worth reading. Here are a few:

1. ANSI Standard Z87.1 *Safe Practices for Occupation and Educational Eye and Face Protection*. American National Standards Institute, 1430 Broadway, New York, NY 10018.

2. NFPA Standard 70-1978, *National Electric Code*, National Fire Protection Association, 470 Atlantic Avenue, Boston, MA 02210.

3. OSHA, *Safety and Health Standards*, 29FR 1910, U.S. Government Printing Office, Washington, D.C. 20402

MAINTENANCE

As with any electrical power device, the electrical connections should be checked periodically for tightness and corrosion. If corrosion exists, clean the connection and then tighten. If the manufacturer provides torque specifications, then these specifications should be followed. If the information is not provided, then common sense should prevail.

As mentioned in previous chapters, the life of all electrical equipment is greatly affected by the overall operating temperature of the power semi-conductors, the electronic circuit boards and all electrical/mechanical connections.

If the Maintenance Manual for the equipment is not available then refer to the chart in Figure 37 for the proper torque.

Screw/Bolt Size	Torque Value
6-32	Tight to the feel
8-32	15 inch-pounds
10-32	30 inch-pounds
1/4-20	70 inch-pounds
5/16-18	100 inch-pounds
3/8-16	375 inch-pounds
1/2-13	500 inch-pounds

Figure 37 - Torque Ratings Chart

If a mechanical lug such as those manufactured by Burndy Corporation or ILSCO is used, check the manufacturer's data sheet for the tightness information.

Periodically check the insulation on all input and output cables. Failure to do so can result in poor starting, equipment failure and/or personnel harm.

SUMMARY

Safety is of the utmost importance. Eyes and skin can be harmed by the intense ultraviolet and infrared light produced by the plasma arc. Hearing can be harmed by the loud noise associated with plasma arcs. Always operate the equipment in a safe manner including the use of proper safety equipment.

Periodic inspections of electrical connections are important to the overall life of plasma equipment.

NOTES CHAPTER 13

PLASMA INDUSTRY CONTACTS
& INDUSTRIAL ASSOCIATIONS
(In alphabetical order)

AMETEK HDR Power Systems
Dan Skinner, Business Manager
3563 Interchange Rd.
Columbus, Ohio 43204
(614) 308-5500
http://www.hdrpower.com

Briggs Machining Co.
23190 Del Lago
Laguna Hills, CA 92653
(949)770-1160

Diversified Controls and Systems, Inc.
Joe Purcell, President
645 Persons St.
East Aurora, NY 14052
(716) 652-1460
http://www.diversifiedcontrols.com

Encore Environmental Solutions, Inc.
Gary Von Sesen, CEO
932 Heykoop Drive
Morristown, TN 37814
(423) 587-7383
http://www.encoreenvironmental.com

Geoplasma,LLC
171 17th St. NW. Suite 1550
Atlanta, GA 30363
(678) 538-4321
http://www.geoplasma.com

Hardface Alloys Inc.
Daren Gansert
9230 Norwalk Blvd.
Santa Fe Springs, CA 90670
(562) 463-8133

HI Disposal Systems
PO Box 1724
Indianapolis, IN 46206
(866) 500-1724
http://www.hawkinsindustries.com

Hitachi Metals
Environmental Systems Company
+81-3-5765-4701
http://www.hitachi-metals.co.jp/e/prod/prod07/p07_2_02.html

Hypertherm, Inc.
Joel Kine
PO Box 5010
Etna Road
Hanover, NH 03755
(603) 643-3441
http://www.hypertherm.com

Integrated Environmental
David Lamar
1935 Butler Loop
Richland, WA 99352
(509) 946-5700
http://www.inentec.com

Macura Instruments
Pavel Macura
140.00 Praha 4
Czech Republic
+420 (0) 2 41402417
macura@comp.cz

MSE Technology Applications, Inc.
200 Technology Way
PO Box 4087
Butte, MT 59702
(406) 494-7100
contact@msw.com

Oxilume LLC
Todd Foret
252 McCarty Rd.
Houston, TX 77029
(832) 250-2982
(337) 298-8870
http://www.oxilume.com

PEAT, Inc.
Martin Springer, President
Huntsville, AL
(256) 859-3006
http://www.peat.com

Phoenix Solutions Company
Douglas Frame, President
3324 Winpark Drive
Crystal, MN 55427
(763) 544-2721
http://www.phoenixsolutionsco.com

Plasma Environmental Technologies, Inc.
130 Adelaide Street West, Suite 2320
Toronto, Ontario, CANADA M5H3P5
(416) 599-9979
http://www.plasmaenvironmental.com

Praxair, Inc.
Worldwide Headquarters
39 Old Ridgebury Rd.
Danbury, CT 06810
1-800-772-9985
http://www.praxair.com

Progressive Technologies, Inc.
Doug Kwaiser
4695 Danvers Dr. SE
Grand Rapids, MI 49512
(616) 285-8312
http://www.ptihome.com

PyroGenesis
Luigi Sperduti, Philippe Chevalier
2000 Rue William
Montreal (Quebec)
Canada H3J 1R8
(514) 937-0002
http://www.pyrogensis.com

Resorption Canada Ltd.
Randy Bennett – Director of Business Development
2610 Del Zotto Avenue
Gloucester, Ontario, CANADA K1T3V7
http://www.rcl-plasma.com

Shaw Environmental & Infrastructure, Inc.
312 Directors Dr.
Knoxville, TN 37923
(865) 690-3211
http://wwwshawgrp.com

Solena Group (The)
Dennis Miller, Chief Scientist
Ronald Reagan Building and Intl Trade Center
1300 Pennsylvania Ave.
Washnington D.C. 20004
(202) 682-2405
http://www.solenagroup.com

SRL Plasma Ltd.
Rex Williams
PO Box 119
Narangba, Queensland
AUSTRALIA 4504
617-3203-3400
http://www.sriplasma.com

Sulzer-Metco AG
Rigackerstr.16
CH-5610 Wohlan
Switzerland
+41 56 618 81 91
http://www.sulzermetco.com

Tafa-Praxair Surface Technologies
Jonathan Head
1500 Polco St.
Indianapolis, IN 46224
(317) 240-2500
http://www.praxair.com/thermalspray

Tetronics Limited
David Deegan
Wicklesham Farm
Faringdon, Oxon
UNITED KINGDOM SN7 7PN
+44 (0) 1367 240224
http://www.tetronics.com

Vanguard Research, Inc.
10400 Easton Place Suite 450
Fairfax, VA 22030
(703) 934-6300
http://www.jdmag.wpafb.af.mil/peps.pdf

Weldstar
Dan Regnier
1750 Mitchell Rd.
PO Box 1150
Aurora, IL 60507
(630) 859-3100
http://www.weldstar.com

Wayne Industrial Tech Services, LTD.
Wayne Cox
7887 Ohio Northern Dr. W
Lancaster, OH 43130
(614) 862-9487
http://www.witsltd.us

Westinghouse Plasma Corporation
Shyam Dighe
Plasma Center – Waltz Mill Site
Madison, PA 15663
(724) 722-7052
http://www.westinghouse-plasma.com

INDUSTRIAL ASSOCIATIONS

AMERICAN WELDING SOCIETY – AWS

ASSOCIATION CONNECTING ELECTRONICS INDUSTRIES – IPC

ASSOCIATION for MANUFACTURING EXCELLENCE – AME

ASSOCIATION for MANUFACTURING TECHNOLOGY – AMT

CANADIAN ELECTRIC CODE - CEC

CANADIAN WELDING ASSOCIATION – CWA

GASES and WELDING DISTRIBUTORS ASSOCIATION – GAWDA

INTERNATIONAL ELECTRIC CODE – IEC

INTERNATION THERMAL SPRAY ASSOCIATION - ITSM

NATIONAL ASSOCIATION of MANUFACTURERS – NAM

NATIONAL ELECTRIC CODE - NEC

NATIONAL ELECTRICAL MANUFACTURERS ASSOCIATION – NEMA

SOCIETY of MANUFACTURING ENGINEERS – SME

WELDING EQUIPMENT MANUFACTURERS COMMITTEE – WEMCO

GLOSSARY OF TERMS

AC – Alternating Current. The power source at the electrical outlet in your home is AC.

AMBIENT TEMPERATURE – this is the temperature in which an SCR Power Control is expected to operate in. This is usually 0☐ to 50☐.

ANODE - the negative power terminal of an SCR or Diode.

CATHODE - the positive power terminal of an SCR or Diode.

COMMAND SIGNAL - a variable input to a Plasma Power Supply that determines the output current setting. It can be a current, voltage or potentiometer input.

CSA – This stands for Canadian Standards Association. This organization provides 3rd party certifications of a product to Canadian Standards.

cUL - Underwriters Laboratories' mark for 3rd party certification of a product to Canadian Standards.

CUPOLA – A type of furnace used for plasma heating.

CURRENT REGULATION - the means for regulating the current to a ever-changing load resistance. It also linearizes the input command signal to the output current.

DC – Direct Current. A battery produces DC. DC is used to control the arc in a plasma application.

DC POWER SUPPLY – An electrical device that converts the utility's AC power to DC current for controlling the plasma arc.

DIODE - a semiconductor that allows current to flow in only one direction. It has no control terminal.

ENCLOSURE - the "box" in which a Plasma Power Supply is contained in - Usually a NEMA type such as a NEMA 1

FURNACE – A containment area where the plasma torch performs it's work.

HARMONICS - are generated when electrical energy is switched and is normally undesirable. Harmonics are expressed as a multiple of the fundamental frequency.

LED – this stands for Light Emitting Diode. Typically used as diagnostic indicators on electronic equipment.

MSW – "Municipal Solid Waste", the waste normally found in the dumps and/or landfills.

ONO-TRANSFERRED TORCH – The arc is between two electrodes inside the plasma torch.

PLASMA – A naturally occurring phenomenon such as lightning. Commonly referred to as the "forth state of matter".

PLASMA CUTTING – A method using the high temperatures produced by a plasma torch for cutting metals.

PLASMA SPRAYING – A method used for powder coating using high temperatures produced by a plasma torch.

PLASMA TORCH – A device that delivers the plasma heat to the load.

POWER FACTOR (P.F.) - the utilization factor of the power being consumed from the power source. A poor power factor means the user is paying for energy not being used to produce heat for example.

PWM – "Pulse-Width-Modulated" a method of controlling power. The pulse is the "on/off" action and the width is the "time-on vs time-off" ratio.

PYROLYSIS – The process of causing a chemical change using heat in an oxygen-starved environment.

R-C SNUBBER - series connected resistor and capacitor network connected across an SCR to "slow down" the rate of applied voltage (dv/dt) to help prevent the SCR from falsely turning on. This is used in conjunction with a MOV for maximum protection.

RADIO FREQUENCY INTERFERENCE (RFI) – this is high frequency electrical interference generated by the chopping action of phase-fired SCRs.

RECTIFIER – Solid-State device for converting Alternating Current (AC) to Direct Current (DC).

SCR - Silicon Controlled Rectifier, a device used to switch the power to the load in an SCR Power Control

SEMICONDUCTOR FUSE - the protective device used to save an SCR in the case of an overcurrent condition. It is usually a very fast acting fuse. Not for wire/cable protection.

THYRISTOR - see SCR.

TRANSFERRED ARC – The arc is transferred from inside the plasma torch to the work-piece. The torch is one electrode, the work-piece is the other.

UL - Underwriters Laboratories. This organization provides 3rd party certification of products. The mark indicates the product meets the stringent safety specifications. It is commonly required for installation of electrical equipment in large cities such as Chicago or New York City.

VITRIFICATION – A method of "melting" inorganic waste creating a non-toxic material.

BIBLIOGRAPHY

"Power Converter Handbook"
(Canadian General Electric Company Limited)

"Power Electronics and AC Drives"
B.K. Bose.
(Prentice-Hall, Englewood Cliffs, New Jersey 07632)

"SCR Manual - 5th edition"
(General Electric Co., Auburn, N.Y., 1972)

"IEEE Std 519-1992 - IEEE Recommended Practices and Requirements for Harmonic Control in Electrical Power Systems"
(Institute of Electrical and Electronics Engineers, Inc,
New York, N.Y. 10017)

"Spraytime"
(ITSM - First Quarter 2005)

"Spraytime"
(ITSM   Forth Quarter 2004)

"Spraytime"
(ITSM – Septermber 2004)

"H401 & H601 Power Supplies Instruction Manual #800410 Rev 5"
(Hypertherm, Inc. Copyright 2002)

SILICON CONTROLLED RECTIFIER DESIGNERS HANDBOOK
(Westinghouse Electric Corp. Youngwood, Pennsylvania, 1970)

ELECTRIC PROCESS HEATING
Maurice Orfeuil, (Battelle Press, copyright 1982)

AN INTRODUCTION TO SCR POWER CONTROLS
(George A Sites, Copyright 2005)

PLASMA ARC HEATING – AN OVERVIEW
(George A Sites, Industrial Heating – May 1991)

IMPROVING PLASMA SPRAY COATING CONTROL
(George A Sites, Process Heating – April 2005)

THE WALL STREET JOURNAL on line
(Jim Carlton, February 17, 2005)

INDEX

non-transferred, xxiii, 11, 37, 40

O

operating, 15, 19, 57, 60, 82, 83, 84, 85, 97

OSHA, 5, 97

Oxilume, C

P

PCBs, 30, 32

PEAT, 31, C

PEC, 27

Phoenix Solutions, 20, 21, 27, 29, C

plasma, i, iii, v, xi, xv, xvii, xix, xxi, xxiii, xxv, 1, 3, 4, 5, 9, 10, 11, 19, 20, 22, 24, 27, 28, 29, 31, 32, 33, 34, 37, 39, 40, 43, 44, 45, 46, 47, 51, 52, 53, 55, 57, 58, 63, 73, 74, 76, 80, 81, 84, 87, 96, A, D, E, F, I, II, b

Plasma, xi, xv, xvii, xxiii, xxv, 3, 4, 5, 9, 10, 11, 20, 22, 24, 27, 28, 29, 31, 32, 33, 34, 37, 39, 40, 43, 44, 45, 46, 47, 52, 53, 57, 58, 63, 74, 76, 80, 81, 84, 96, D, E, F, I, II

plasma power supply, 12, 39, 58, 59, 63, 68, 69, 73, 74, 76, 84, 85, I, II

plasma torch, 5, 10, 11, 12, 13, 14, 15, 21, 22, 23, 24, 28, 29, 33, 39, 45, 48, 66, 68, 69, 88, 90, 96, II, III

power factor, xix, 58, 59, 76, 82, 83, 85, II

Praxair, 40, D

process, 5, 6, 12, 13, 14, 19, 20, 22, 27, 28, 29, 31, 32, 34, 37, 41, 43, 48, 69, 70, 74, 82, 85, 88, II

Progressive Technologies, D

PWM, xix, xxv, 59, 60, 73, 74, 75, 76, 77, 81, 82, 84, 85, II

www.ingramcontent.com/pod-product-compliance
Lightning Source LLC
Chambersburg PA
CBHW081454170526
45166CB00008B/2424